Quercetin - Effects on Human Health

Edited by Joško Osredkar

Published in London, United Kingdom

Quercetin - Effects on Human Health
http://dx.doi.org/10.5772/intechopen.111270
Edited by Joško Osredkar

Contributors
Abdallah Z. Zayan, Ahmed M. Sayed, Aimi Syamima Abd Manap, Bernadette Xin Jie Tune, Brendan D. Stamper, Falah H. Shari, Joško Osredkar, Kalaivani Batumalaie, Kate Nyarko, Ker Woon Choy, Mahendran Sekar, Marek Pytliak, Mazhar Mushtaq, Mingrui Li, Mohammed I. Rushdi, Mosad A. Ghareeb, Muhammad Candragupta Jihwaprani, Neeraj Kumar Fuloria, Nhung Au, Priya Mijgar, Rhanye Mac Guad, Sheryar Afzal, Shivkanya Fuloria, Uday Deokate, Vetriselvan Subramaniyan, Viliam Vaník, Wahyu Choirur Rizky, Yuan Seng Wu, Yun Gao

First published in London, United Kingdom, 2024 by IntechOpen
IntechOpen is the global imprint of INTECHOPEN LIMITED, registered in England and Wales, registration number: 11086078, 167-169 Great Portland Street, London, W1W 5PF, United Kingdom

British Library Cataloguing-in-Publication Data
A catalogue record for this book is available from the British Library

Additional hard and PDF copies can be obtained from orders@intechopen.com

Quercetin - Effects on Human Health
Edited by Joško Osredkar
p. cm.
Print ISBN 978-0-85466-527-3
Online ISBN 978-0-85466-526-6
eBook (PDF) ISBN 978-0-85466-528-0

For EU product safety concerns:
IN TECH d.o.o., Prolaz Marije Krucifikse Kozulić 3, 51000 Rijeka, Croatia,
info@intechopen.com or visit our website at intechopen.com.

Meet the editor

Joško Osredkar began his career at the University Medical Centre Ljubljana, Slovenia, in 1979. In 1987, he started his specialization in medical biochemistry, earning a Ph.D. in 1992 and becoming a Full Professor of Clinical Biochemistry in 2005. His recent research focuses on oxidative stress in the developing fetus and early childhood. Dr. Osredkar has been acknowledged for his proficiency in doping management, becoming a member of the Doping Control Expert Group of the World Anti-Doping Agency (WADA) in 2003. He was granted a special honor by the International Olympic Committee (IOC) in 2009. In 2018, he received a nomination to the board of directors of the International Society of Oncology and Biomarkers (ISOBM) and chaired the 46th ISOBM Congress as part of this commitment.

Contents

Preface

Quercetin is a flavonoid prevalent in fruits and vegetables that has been a staple in human diets for generations. This book provides a comprehensive overview of quercetin in six sections.

Section 1 highlights key points about the effects of quercetin on human health. It also addresses quercetin's many health advantages, including its antiviral, antibacterial, anti-inflammatory, antioxidant, and anticancer qualities.

Section 2 discusses how vital it is to comprehend quercetin's pharmacokinetics in order to evaluate its possible therapeutic uses. Because of its wide spectrum of biological activity, quercetin is a key pharmacological issue. Quercetin has demonstrated potential uses in the management and avoidance of several illnesses, such as inflammatory disorders, allergies, metabolic problems, and eye diseases.

Section 3 presents the complexity of the molecular pathways of quercetin, which exhibit variability based on the particular cell type and situation. In addition to its demonstrated ability to alter the molecular processes and signaling pathways linked to inflammation, quercetin has been studied for possible anticancer properties. It can also go through several metabolic processes in the body, such as O-methylation, sulfation, and glucuronidation, or it can modify molecular events and signaling pathways involved in cell survival, proliferation, differentiation, angiogenesis, invasion, migration, and metastasis.

Section 4 explores quercetin's potential uses in human health and status as an antioxidant. Due to the presence of four hydroxyl groups on the benzo-dihydropyran ring in its chemical structure, quercetin possesses a potent antioxidant ability. Quercetin can thus help the body maintain a steady state and eliminate free radicals.

Section 5 presents the possible therapeutic benefits of quercetin as a promising option for oxidative stress and inflammatory diseases. There is potential for quercetin to help prevent and treat cancer. Its capacity to cause apoptosis or cell death, prevent the proliferation of cancer cells, and decrease the chances of developing specific cancer types have been investigated. Research has been done on quercetin's antiviral qualities, which may include preventing the spread of viruses, as well as its possible advantages in treating metabolic conditions including diabetes and obesity. It might aid in lowering inflammation, enhancing insulin sensitivity, and controlling glucose metabolism. The potential therapeutic benefits of quercetin have also been investigated in relation to gastrointestinal issues, skin ailments, arthritic conditions, and other conditions.

Section 6 outlines studies on quercetin's potential positive effects on the cardiovascular system. Supplementing with quercetin has been demonstrated to lower systolic blood pressure in people with type 2 diabetes, suggesting a possible antihypertensive impact. In terms of cardioprotection, quercetin has been linked to enhanced endothelial function, which is critical for keeping blood arteries in good condition. Further research shown that quercetin might have preventive effects on the development of atherosclerosis. In addition, quercetin has demonstrated advantages in enhancing heart performance in animal models of post-ischemic damage and pressure overload.

Joško Osredkar
Clinical Institute for Clinical Chemistry and Biochemistry,
University Medical Centre Ljubljana,
Ljubljana, Slovenia

Section 1

Introduction and Overview

Chapter 1

Introductory Chapter: Quercetin – Effects on Human Health

Joško Osredkar

1. Introduction

Quercetin is a flavonoid compound found in a variety of fruits, vegetables, and grains. It is known for its antioxidant and anti-inflammatory properties and has been studied extensively for its potential health benefits. The main issues related to quercetin and its effects on human health are antioxidant properties, anti-inflammatory properties, immune system support, cardiovascular health, allergy relief, cancer prevention, and brain health.

Strong antioxidant qualities of quercetin have been demonstrated, suggesting that it may help shield cells from harm brought on by free radicals. This could aid in the prevention of long-term conditions like cancer, heart disease, and neurological illnesses. Additionally, quercetin possesses anti-inflammatory qualities that may aid in lowering inflammation all over the body. This could lessen the signs and symptoms of inflammatory diseases such as inflammatory bowel disease and rheumatoid arthritis. It has been demonstrated that quercetin supports the immune system by lowering inflammation and stimulating immune cell function. This might facilitate better disease recovery and help avoid infections. Quercetin has the potential to enhance cardiovascular health by mitigating oxidative stress and inflammation, two factors that increase the risk of heart disease. Additionally, it might lower blood pressure and enhance blood flow. Due to its demonstrated antihistamine qualities, quercetin may be able to lessen allergy symptoms like hives, runny nose, and itchy eyes. Quercetin may have anticancer effects and be able to stop the growth and spread of cancer cells, according to certain research. Because of its potential neuroprotective qualities, quercetin may enhance cognitive performance and lower the risk of neurodegenerative illnesses like Parkinson's and Alzheimer's [1].

2. Signaling pathways associated with quercetin's effects on health

Antioxidant defense pathway
The Nrf2 (nuclear factor erythroid 2-related factor 2) signaling pathway
NF-kB (nuclear factor-kappa B) pathway
Apoptotic (cell death) pathways
Insulin signaling pathway
Mitogen-activated protein kinase (MAPK) pathways

IntechOpen

Wnt/β-catenin pathway
AMP-activated protein kinase (AMPK)
Hepatitis C virus (HCV) pathway.

3. Quercetin, as a flavonoid antioxidant

It is well known that quercetin has antioxidant qualities. Antioxidants are substances that assist in shielding cells against reactive oxygen species (ROS) and free radicals, which can cause oxidative damage. Cell damage and inflammation are two health problems that can be attributed to oxidative stress, which is the result of an imbalance between the body's capacity to eliminate free radicals and their creation. Quercetin has the ability to combat free radicals, boost the body's natural antioxidant defense, lessen inflammation, guard against chronic illnesses, and support healthy immune system function.

4. Absorption of quercetin

The absorption of quercetin can be influenced by various substances, both enhancers and inhibitors. Quercetin is known to have limited bioavailability when consumed orally, meaning that a significant portion of it may not be efficiently absorbed into the bloodstream. Substances and factors that can affect the absorption of quercetin:

Piperine (black pepper extract), vitamin C (ascorbic acid), bromelain (pineapple enzyme), fats and oils, quercetin glycosides, and heat and cooking are factors that improve the absorption of quercetin.

5. Impact on biochemical parameters

The possible effects of quercetin on a number of bodily biochemical markers have been investigated. Quercetin's antioxidant, anti-inflammatory, and metabolic regulating qualities are responsible for some of the effects on biochemical outcomes that have been documented. The biochemical effects of quercetin, along with the mechanisms underlying them, include blood pressure regulation, glucose and insulin metabolism, lipid metabolism, antioxidant activity, and anti-inflammatory actions.

Strong antioxidant quercetin has the ability to counteract free radicals and lessen oxidative stress. Numerous biochemical effects, including decreased oxidative damage and improved lipid profile, can result from this antioxidant action. Reactive oxygen species (ROS) can cause oxidative damage to biological components such as proteins, lipids, and DNA. Quercetin can help prevent this damage. Lower levels of oxidative stress indicators in the body may arise from this. One of the main contributors to the development of atherosclerosis is oxidized LDL cholesterol, which quercetin may help lower levels of. A better lipid profile may result from reducing oxidized LDL [2].

Because of its anti-inflammatory qualities, quercetin has been shown to lower inflammatory cytokines and reduce C-reactive protein (CRP). Pro-inflammatory cytokines like interleukin-6 (IL-6) and tumor necrosis factor-alpha (TNF-α) can also be produced and released less frequently when quercetin is present. As a result, systemic inflammation is reduced. CRP is a measure of systemic inflammation, and some research has linked quercetin consumption to decreased CRP levels [3].

Quercetin may have biochemical effects related to blood pressure control. Vasodilation–Quercetin can promote the relaxation of blood vessels (vasodilation), which can lead to improved blood flow and lower blood pressure. Endothelial function—quercetin may enhance endothelial function by increasing the production of nitric oxide (NO), a molecule that helps regulate blood vessel tone.

Quercetin may have an effect on biochemical markers associated with glucose and insulin metabolism. It may improve insulin sensitivity in cells, which may result in better glucose regulation and lower fasting insulin levels. Additionally, some studies have shown that quercetin supplementation may lower glycosylated hemoglobin (HbA1c), a long-term indicator of blood glucose control.

Reduced triglycerides are one of the biochemical indicators that reveal how quercetin affects lipid metabolism. Triglyceride levels can be lowered by quercetin, which is beneficial for cardiovascular health. Liver function may be affected by quercetin's biochemical activities. Supplementing with quercetin has occasionally been linked to lower liver enzyme levels, indicating possible enhancements to liver function.

6. Quercetin plays an important role in the treatment of human diseases through epigenetic mechanisms

Because of its possible application in the management of human illnesses, quercetin has drawn interest from the epigenetics community. The term "epigenetics" describes modifications to gene expression that can have a major effect on a person's health but do not entail changes to the DNA sequence. There are numerous ways that quercetin is thought to affect epigenetic pathways. Through epigenetic pathways, quercetin may be useful in treating the following diseases [4–6]:

DNA methylation: quercetin has been studied for its ability to modulate DNA methylation, an epigenetic modification that involves the addition of methyl groups to DNA molecules. DNA methylation patterns can influence gene expression. Quercetin may have demethylating effects on certain genes, potentially leading to changes in gene expression patterns that are relevant to disease prevention or treatment.

Histone modification: quercetin may impact histone modifications, such as acetylation and methylation. Histones are proteins that help package DNA into a compact structure. Alterations in histone modifications can affect chromatin structure and gene accessibility. Quercetin's influence on histone modifications may lead to changes in gene expression associated with disease pathways.

MicroRNA regulation: quercetin has been shown to regulate the expression of microRNAs (miRNAs), small non-coding RNA molecules that play a role in post-transcriptional gene regulation. Changes in miRNA expression can have epigenetic effects by modulating the stability and translation of messenger RNAs (mRNAs). Quercetin's ability to influence miRNAs may impact disease-related pathways.

Anti-inflammatory effects: quercetin's epigenetic effects may extend to its anti-inflammatory properties. Chronic inflammation is a key factor in the development of many diseases, and quercetin's ability to modulate inflammatory gene expression may contribute to its therapeutic potential.

Antioxidant and DNA repair: by reducing oxidative stress and promoting DNA repair mechanisms, quercetin may indirectly influence epigenetic stability. Oxidative stress can lead to epigenetic changes, and quercetin's antioxidant properties may mitigate these effects.

Cancer: quercetin has been studied for its potential role in cancer prevention and treatment. Epigenetic changes, such as DNA methylation and histone modifications, are common in cancer. Quercetin's ability to modulate epigenetic mechanisms may influence cancer-related gene expression and pathways.

7. Antimicrobical activity of quercetin

Quercetin shows antimicrobial activity against a wide range of microorganisms, including bacteria, viruses, fungi, and parasites. Several mechanisms are involved in quercetin's antimicrobial activity:

Disruption of cell membranes
Inhibition of enzyme activity
DNA and RNA interactions
Antioxidant activity
Interference with viral replication
Modulation of immune responses
Biofilm disruption.

The effectiveness of quercetin as an antimicrobial agent can vary depending on factors such as the specific microorganism, its susceptibility to quercetin, and the concentration and delivery method of quercetin used. While quercetin shows promise as a natural antimicrobial compound, more research is needed, including clinical studies, to better understand its potential clinical applications and optimal use in the treatment or prevention of microbial infections.

8. Quercetin and cardiovascular diseases

Quercetin has been studied for its potential cardiovascular benefits. Quercetin is considered a cardiovascular agent due to several potential cardioprotective properties:

Antioxidant activity
Anti-inflammatory effects
Blood pressure regulation
Endothelial function
Cholesterol management
Platelet aggregation inhibition
Nitric oxide production
Cardiometabolic health.

9. Wound healing

Quercetin has been studied for its potential role in wound healing. Its mechanisms of action in wound healing are complex and multifaceted:

Anti-inflammatory effects
Antioxidant activity
Collagen synthesis
Angiogenesis promotion
Fibroblast proliferation and migration
Epithelialization
Antimicrobial activity
Scar reduction.

The specific mechanisms underlying quercetin's effects on wound healing may vary depending on the type and severity of the wound. Quercetin can be applied topically to the skin for its antioxidant, anti-inflammatory, and potential UV-protective properties. To enhance the delivery and effectiveness of topical quercetin applications, various delivery strategies and formulations can be considered:

Creams and lotions
Gels
Serums
Nanoparticles and liposomes
Microneedling
Transdermal patches
Sprays
Lip balms
Sunscreen formulations.

Quercetin has been studied for its potential to contribute to skin brightening and improvement of skin complexion. Skin brightening typically refers to the reduction of hyperpigmentation, such as dark spots, uneven skin tone, and age-related skin discolorations. Quercetin may be associated with skin brightening:

Antioxidant properties
Anti-inflammatory effects
Inhibition of tyrosinase
Sun damage protection
Collagen production
Reduction of red and brown spots
Antioxidant synergy.

10. Skin cancer

The development of skin cancer, particularly melanoma, is primarily driven by exposure to ultraviolet (UV) radiation from the sun or artificial sources like tanning beds. UV radiation causes DNA damage in skin cells, leading to mutations that can trigger the development of skin cancer. Quercetin has been studied for its potential role in protecting against UV-induced skin damage and, to some extent, in mitigating the mechanisms underlying skin cancer. Quercetin may be connected to the mechanisms of skin cancer:

Antioxidant activity
Anti-inflammatory effects

DNA repair
Protection against UV-induced immunomodulation
Inhibition of cell proliferation
Melanin production regulation.

Skin cancer prevention primarily involves sun protection measures, such as wearing sunscreen, protective clothing, and sunglasses, seeking shade, and avoiding tanning beds. Quercetin-containing supplements or topical products should be used cautiously and under the guidance of a healthcare professional, especially in the context of skin cancer prevention or treatment. They should not be considered a sole substitute for established sun protection practices or medical treatment.

11. Quercetin in nanoformulations

The use of quercetin in nanoformulations presents both opportunities and challenges in the quest for improved clinical efficacy and enhanced biological effects. Nanoformulations of quercetin involve the creation of nanoscale structures (nanoparticles, liposomes, micelles, etc.) to encapsulate or deliver quercetin to target tissues or cells. While these approaches have shown promise in enhancing the therapeutic potential of quercetin, they also pose certain challenges.

11.1 Opportunities

Improved bioavailability: quercetin has limited bioavailability when consumed orally, as it can be rapidly metabolized and excreted. Nanoformulations can protect quercetin from degradation and enhance its absorption, leading to higher levels in the bloodstream and target tissues.

Targeted delivery: nanoformulations can be engineered to deliver quercetin specifically to the intended site of action, such as tumor tissues or inflamed areas. This targeted delivery can minimize off-target effects and improve therapeutic outcomes.

Sustained release: some nanoformulations allow for controlled and sustained release of quercetin over an extended period. This can maintain therapeutic levels of quercetin in the body and reduce the frequency of dosing.

Combination therapies: quercetin-loaded nanoparticles can be combined with other drugs or therapeutic agents to create synergistic effects. This approach is particularly relevant in cancer therapy, where quercetin may enhance the efficacy of chemotherapy or radiation therapy.

Reduced side effects: by improving the selectivity of quercetin delivery, nanoformulations may reduce the risk of side effects associated with systemic quercetin administration.

11.2 Challenges

Complex formulation development: designing effective quercetin nanoformulations can be complex and require expertise in nanotechnology. The choice of nanocarrier, formulation conditions, and stability considerations must be carefully addressed.

Regulatory approval: nanoformulations often face regulatory challenges due to their unique characteristics. Ensuring their safety and efficacy through rigorous testing and clinical trials can be time-consuming and costly.

Scale-up and manufacturing: scaling up the production of quercetin nanoformulations for commercial use can be challenging and may affect the consistency and cost-effectiveness of manufacturing.

Biocompatibility: the biocompatibility of nanomaterials used in quercetin nanoformulations is a critical concern. Ensuring that the materials are safe for use in humans is essential.

Clinical translation: moving from preclinical studies to clinical trials with nanoformulations can be a significant hurdle. Demonstrating improved clinical efficacy compared to conventional quercetin formulations is essential for their acceptance in clinical practice.

Cost: developing and producing quercetin nanoformulations can be expensive, which may affect their affordability and accessibility to patients.

12. Quercetin supplementation

Quercetin is generally considered safe when consumed as part of a normal diet. However, when using quercetin supplements at high doses or in concentrated forms, there is potential for interactions with medications, including those used for hypertension (high blood pressure) treatment. Some important considerations [7]:

Potentiation of medication effects: quercetin may have vasodilatory effects, which can relax and widen blood vessels. While this effect might be beneficial in some cases, it can also potentiate the effects of antihypertensive medications, potentially leading to excessively low blood pressure (hypotension).

Interaction with blood pressure medications: quercetin might interact with certain antihypertensive medications, such as calcium channel blockers, ACE inhibitors, and beta-blockers. Combining quercetin supplements with these medications may enhance their blood pressure-lowering effects, which could result in hypotension. It is essential to monitor blood pressure closely in such cases.

Effect on drug metabolism: some studies suggest that quercetin may inhibit cytochrome P450 enzymes, which are involved in drug metabolism. This inhibition could potentially affect the metabolism and clearance of certain medications, potentially leading to altered drug levels in the body.

Potential interaction with diuretics: quercetin is found in foods like onions, apples, and tea, and it has mild diuretic properties. When combined with diuretic medications, quercetin's diuretic effects might be additive, potentially leading to an increased risk of dehydration or electrolyte imbalances.

Individual variability: interactions between quercetin and hypertension medications can vary among individuals. Some people may be more sensitive to these interactions than others, and factors such as the specific medication, dosage, and overall health status can influence the outcome.

13. Conclusion

Quercetin is a flavonoid compound found in various fruits, vegetables, and grains, known for its antioxidant, anti-inflammatory, and potential therapeutic properties.

13.1 Mode of action

Antioxidant activity: quercetin acts as a potent antioxidant, scavenging free radicals and reducing oxidative stress in the body. This antioxidant activity helps protect cells and tissues from damage caused by reactive oxygen species (ROS).

Anti-inflammatory effects: quercetin has anti-inflammatory properties that can help reduce inflammation by inhibiting inflammatory mediators and pathways. It may modulate the activity of cytokines, enzymes, and transcription factors involved in the inflammatory response.

Modulation of cellular signaling: quercetin can influence various cellular signaling pathways involved in processes such as cell proliferation, apoptosis, and gene expression. These effects may contribute to its therapeutic potential in different diseases.

13.2 Use in different diseases

Cardiovascular health: quercetin has been studied for its potential cardiovascular benefits, including its ability to improve endothelial function, reduce oxidative stress, and lower blood pressure and cholesterol levels. It may also help protect against atherosclerosis and other cardiovascular diseases.

Immune function: quercetin has immunomodulatory effects that may enhance immune function and reduce inflammation. It has been investigated for its potential in supporting immune health and combating infections.

Metabolic health: quercetin may have benefits for metabolic health, including its ability to improve insulin sensitivity, regulate blood glucose levels, and reduce the risk of obesity-related complications.

Cancer prevention: some studies suggest that quercetin may have anticancer properties, including its ability to inhibit tumor cell growth, induce apoptosis (cell death) in cancer cells, and modulate signaling pathways involved in carcinogenesis.

Neurological disorders: quercetin has been investigated for its potential neuroprotective effects in conditions such as Alzheimer's disease, Parkinson's disease, and stroke. Its antioxidant and anti-inflammatory properties may help protect neurons from damage and reduce neuroinflammation.

Skin health: quercetin may have benefits for skin health, including its ability to protect against UV-induced damage, reduce inflammation, and promote wound healing. It has been studied for its potential in treating conditions such as atopic dermatitis, psoriasis, and skin cancer.

13.3 Supplementation

Quercetin supplements are available in various forms, including capsules, tablets, and powders [6].

Dosages of quercetin supplements can vary depending on the specific product and intended use. It is essential to follow the manufacturer's recommended dosage instructions or consult with a healthcare professional for personalized guidance.

Quercetin supplements are often used as part of a broader approach to health and wellness, including a balanced diet, regular exercise, and other lifestyle interventions.

While generally considered safe for most people when taken at recommended dosages, quercetin supplements may interact with certain medications and have potential side effects in some individuals.

In summary, quercetin exhibits diverse biological activities and has been investigated for its potential therapeutic benefits in various diseases and health conditions. While research is ongoing to better understand its mechanisms of action and clinical applications, quercetin supplementation may offer potential benefits for promoting overall health and wellness. As with any supplement, it is important to use quercetin under the guidance of a healthcare professional to ensure safety and efficacy.

Author details

Joško Osredkar
University Medical Centre, Ljubljana, Slovenia

*Address all correspondence to: josko.osredkar@kclj.si

References

[1] Boots AW, Haenen GR, Bast A. Health effects of quercetin: From antioxidant to nutraceutical. European Journal of Pharmacology. 2008;**585**(2-3):325-337. DOI: 10.1016/j.ejphar.2008.03.008

[2] Egert S, Bosy-Westphal A, Seiberl J, et al. Quercetin reduces systolic blood pressure and plasma oxidised low-density lipoprotein concentrations in overweight subjects with a high-cardiovascular disease risk phenotype: A double-blinded, placebo-controlled cross-over study. The British Journal of Nutrition. 2009;**102**(7):1065-1074. DOI: 10.1017/S0007114509371663

[3] Li Y, Yao J, Han C, et al. Quercetin, inflammation and immunity. Nutrients. 2016;**8**(3):167. DOI: 10.3390/nu8030167

[4] Russo M, Spagnuolo C, Tedesco I, et al. The flavonoid quercetin in disease prevention and therapy: Facts and fancies. Biochemical Pharmacology. 2012;**83**(1):6-15. DOI: 10.1016/j.bcp.2011.08.010

[5] Mlcek J, Jurikova T, Skrovankova S, et al. Quercetin and its anti-allergic immune response. Molecules. 2016;**21**(5):623

[6] de Oliveira MR, Nabavi SF, Manayi A, et al. Quercetin and the mitochondria: A mechanistic view. Biotechnology Advances. 2016;**34**(5):532-549

[7] Bagheri R, Asghari G, Farrokhian A, et al. Quercetin supplementation and oxidative stress indices: A systematic review and meta-analysis. Clinical Nutrition ESPEN. 2021;**43**:91-99

Chapter 2

Unveiling the Potential of Quercetin: Chemistry, Health Benefits, Toxicity, and Cutting-Edge Advances

Mosad A. Ghareeb, Abdallah Z. Zayan, Falah H. Shari and Ahmed M. Sayed

Abstract

Quercetin, a naturally occurring flavonoid, has gained significant attention in recent years due to its potential health benefits and versatile applications. This book chapter explores the chemistry of quercetin, shedding light on its molecular structure, biosynthesis, and extraction methods. The chapter delves into the extensive research on the health effects of quercetin, highlighting its antioxidant, anti-inflammatory, anticancer, neuroprotective, and cardioprotective properties. Moreover, the potential risks and toxicity associated with quercetin consumption are thoroughly examined, emphasizing the importance of proper dosage and potential drug interactions. The chapter concludes by providing an overview of recent advances in quercetin development, including nanoformulations, targeted delivery systems, and combination therapies, that hold promise for enhancing its therapeutic efficacy and bioavailability. This comprehensive exploration of quercetin aims to provide researchers, scientists, and healthcare professionals with valuable insights into its multifaceted nature and potential applications in human health.

Keywords: quercetin, flavonoids, health benefits, toxicity, chemistry, antioxidant, anti-inflammatory, anticancer, neuroprotective, cardioprotective, recent advances

1. Introduction

1.1 Overview of quercetin: Definition and significance

Quercetin is a natural compound belonging to the flavonoids family; it is widely distributed in the plant kingdom and has been isolated from many edible and non-edible plants. Quercetin has several pharmacological properties, making it a promising candidate for many pharmaceutical formulations. Nutritionally, quercetin is a major component of the human diet. Onions are the main source of quercetin, in addition to some other plant species such as apples, buckwheat, grapes, tea, cherries, tomatoes, mangoes, citrus, and plums [1, 2].

From the chemical point of view, quercetin is a plant pigment with a flavonol-type structure called 3,3,4,5,7-pentahydroxyflavone and 2-(3,4-dihydroxyphenyl)-3,5,7-trihydroxychromen-4-one) according to IUPAC. It has a molecular formula ($C_{15}H_{10}O_7$) and a molecular weight (302.236 g/mol) [3]. The main structure of the compound is a 15-carbon skeleton, comprising two benzene rings labeled (A-ring and B-ring) connected by a 3-carbon heterocyclic ring labeled (C-ring), in addition to five hydroxyl functional groups [4]. However, the B-ring is considered the main participant in its potent antioxidant activity because it contains a heavy hydroxylation pattern that enables the compound to be a strong free radicals scavenger [5].

On the other side, the compound showed a broad spectrum of biological activities such as anti-SARS-CoV-2, antioxidant, antidiabetic, anticancer, antiarthritic, antiaging, anti-Alzheimer's, antiviral, anti-inflammatory, antimicrobial, anti-allergic, anti-obesity, cardiovascular, hepatoprotective, neuroprotective, and wound-healing [2].

1.2 Historical context and contemporary interest

First identified in 1854 from its glycosidic form, quercitrin, quercetin—named after the Latin word "Quercetum" for Oak Forest—has become a significant compound in pharmaceuticals, cosmetics, and food production [6]. It is predominantly found in a range of health foods and herbal products, with a daily intake ranging from 5 to 40 mg, reflecting its extensive metabolism in the intestine and liver [7, 8]. Quercetin's hydrophilic, stable, and bioavailable characteristics are notably enhanced through glycosylation.

This chapter, "Unveiling the Potential of Quercetin: Chemistry, Health Benefits, Toxicity, and Cutting-Edge Advances," explores quercetin's chemical makeup, therapeutic attributes, and the latest research innovations. It highlights quercetin's role as a cornerstone in the study of plant-based compounds and their impact on human health. Acknowledged for its antioxidant, anti-inflammatory, anticancer, neuroprotective, and cardioprotective properties, quercetin is a ubiquitous dietary component vital for both nutrition and therapy.

The narrative details quercetin's historical progression, its natural synthesis, extraction methods, and the strides made in understanding its complex therapeutic utility. It covers quercetin's chemical structure, emphasizing its distinctive flavonol skeleton and the impact of its hydroxylation on antioxidant activities.

In addressing health implications, the chapter outlines quercetin's mechanisms of action—ranging from disease prevention to managing cardiovascular health, cancer, diabetes, mental health, and neurodegenerative disorders. It also critically assesses the safety and potential risks of quercetin, particularly at high doses, and its interactions with medications, underscoring the importance of cautious supplementation.

Looking to the future, the chapter delves into innovative delivery systems and combination therapies that enhance quercetin's therapeutic potential. It discusses advances in nanotechnology and targeted delivery that aim to improve solubility and bioavailability, positioning quercetin at the forefront of scientific exploration and therapeutic innovation.

Overall, the chapter provides a comprehensive, scholarly examination of quercetin, presenting a balanced view of its benefits and limitations while encouraging further research and development in medicinal chemistry and phytotherapy. It serves as an essential resource for researchers, healthcare professionals, and academicians interested in the intersection of plant-based compounds and health enhancement.

2. Section 1: Quercetin chemistry

2.1 Chemical structure and properties of quercetin

Flavonoids are a group of naturally occurring polyphenolic secondary metabolites that are broadly disseminated in the plant kingdom, particularly in fruits, grains, and vegetables. Flavonoids are also found in the form of aglycones or glycosides. Based on the diversity in their basic structure (C6-C3-C6) regarding the number and location of the hydroxyl groups and the oxidation level of the C-ring, flavonoids could be divided into seven sub-classes namely: flavonols, flavones, isoflavones, anthocyanidins, flavanones, flavanols, and chalcones (**Figure 1**) [9, 10].

Quercetin (3,5,7,3′,4′-pentahydroxyflavone) is a flavonol-type structure, with its IUPAC name verified as [2-(3,4-Dihydroxyphenyl)-3,5,7-trihydroxy-4H-1-benzopyran-4-one] (**Figure 1**). The catechol-type B ring is mainly responsible for most biological activities of the compound; among them are antioxidants [11, 12].

Basically, quercetin occurs in the form of glycosides as well as ethers, sulfates, acyls, and prenyls. The hydroxylation pattern of quercetin (five hydroxy groups) is related to the number of derivatives as well as the biological activities of the compound [13]. Noteworthy, a common location for the glycosylation site is the hydroxyl group at the C-3 position followed by the C-7 position. Quercetin 3-*O*-glycosides exist as mono and/or di-glycosides, and the most common sugar

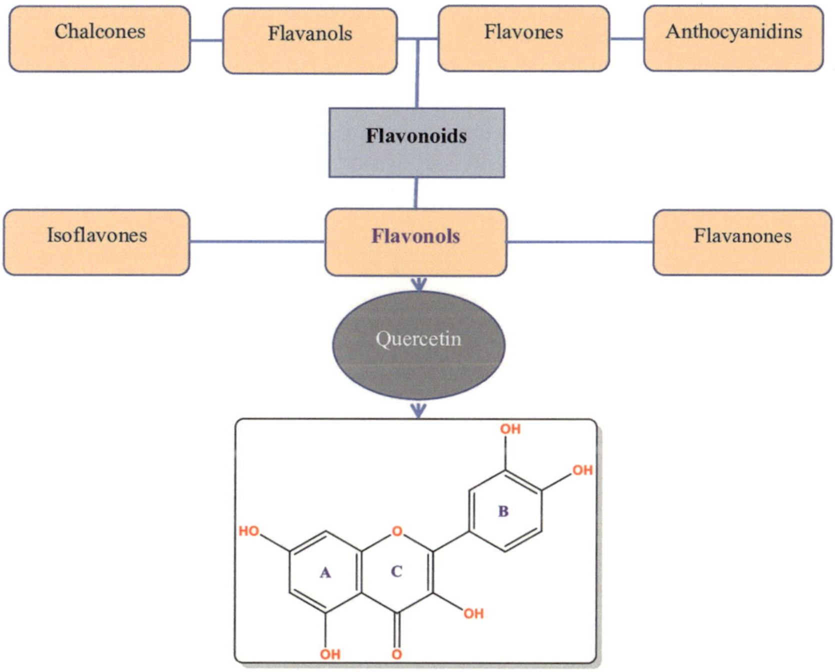

Figure 1.
Chemical classification of flavonoids.

moieties are glucose, galactose, rhamnose, xylose, and glucuronic acid. Quercetin 3-*O*-glycosides include quercetin 3-*O*-glucoside, quercetin 3-*O*-xyloside, quercetin 3-*O*-rhamnoside, quercetin 3-*O*-glucuronide, and quercetin 3-*O*-galactoside as mono-glycosides. In the same context, C-glycosides are likely to occur at the C-6 position [14].

Quercetin is a yellow crystalline solid with a melting point of 316°C and density equal to 1.80 g/cm^3. Moreover, it has a high degree of solubility in organic solvents such as dimethyl sulfoxide (DMSO), dimethyl formamide (DMF), methanol (MeOH), and ethanol (EtOH). On the contrary, it is difficult to absorb quercetin in the body as a hydrophobic compound due to its weak water solubility. The water solubility increases according to the increasing number of sugar moieties in the molecule. Therefore, scientists have turned to conducting many studies in order to improve its water solubility and bioavailability, which reflects positively on its pharmacological properties. Metal chelation is considered one of the methods used to improve its water solubility and bioavailability. However, the most common quercetin-metal complexes are copper and iron complexes [15–17].

2.2 Biosynthesis and natural sources

Plants are exposed to some damages from the surrounding environment, such as UV-radiation and microbial infection, so the process of flavonoids biosynthesis is considered the first line of defense against these external damages [18]. Flavonoids can be biosynthesized via the phenylpropanoid metabolic pathway. This pathway includes a group of enzymes like synthase, isomerases, hydroxylases, and reductases [19]. The biosynthesis of quercetin involves several steps and some enzyme complexes. Firstly, L-Tyrosine or 4-hydroxyphenylalanine is transformed to p-coumaric acid via using tyrosine ammonia lyase (TAL), and then p-coumaric acid is transformed into p-coumaroyl-CoA using 4-coumaroyl-CoA ligase (4CL). Subsequently, p-coumaroyl-CoA reacts with three particles of malonyl-CoA via using chalcone synthase (CHS) which resulted in the development of naringenin chalcone. After that, naringenin chalcone is transformed to naringenin using chalconeisomerase (CHI). Afterward, via using flavanone 3-hydroxylase (F3H) enzyme naringenin is transformed to dihydrokaempferol. Then, dihydrokaempferol is transformed to dihydroquercetin by using flavanone 3′-hydroxylase (F3'H) enzyme. Finally, dihydroquercetin is transformed into quercetin by using flavonol synthase (FLS) enzyme (**Figure 2**) [20].

Quercetin is the most abundant flavonoid, found in fruits, vegetables, and grains. The compound is widely found in many plant species such as apples, honey, raspberries, onions, red grapes, radish, cherries, pears, tomatoes, coriander, green tea, red lettuce, fennel, green beans, asparagus, green pepper, sweet potato, citrus, and dill (**Figure 3**) [18, 21, 22]. As a result of the diversity of the environmental conditions and Ayurvedic culture, the sources of quercetin also vary in different regions around the world [22].

2.3 Methods of extraction and purification

Previous studies indicated that quercetin has been extracted from plant materials using several extraction approaches, such as conventional solvent extraction, ultrasound-assisted extraction (UAE), supercritical fluid extraction (SFE), and microwave-assisted extraction (MAE) [23–26]. Moreover, aqueous mixtures are

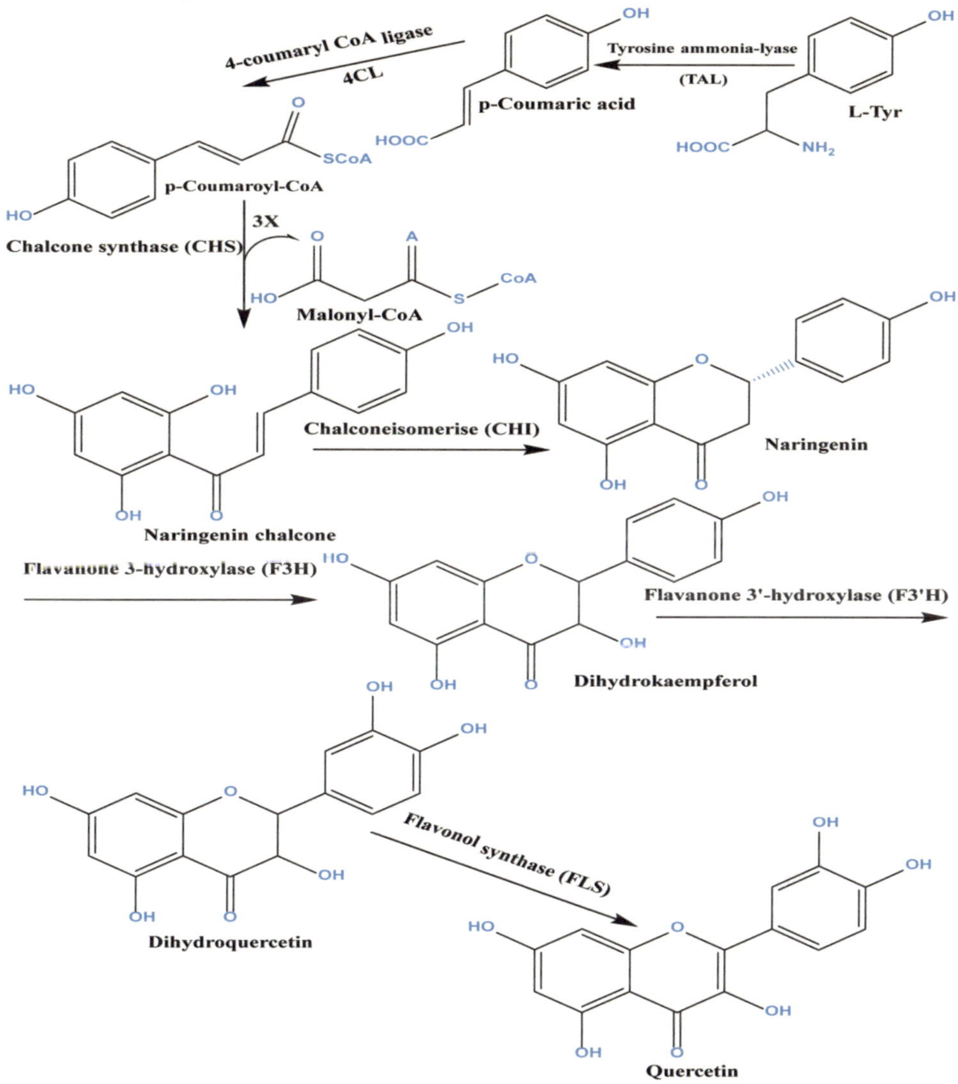

Figure 2.
Biosynthesis pathway of quercetin.

among the best solvents used in extracting quercetin from plant materials in an effective way with sufficient extraction yield, including methanol, ethanol, acetone, and ethyl acetate [27]. For instance, in a previous study quercetin was extracted from dry onion peels with a high extraction yield (21%) using hot aqueous mixture of ethanol (60% ethanol) [28]. Quercetin has been isolated from *Huberantha senjiana* leaf ethyl acetate extract obtained by successive extraction and purified using conventional column chromatography packed with silica gel [29]. In another study, column chromatography and HPLC-UV were applied for the isolation and purification of quercetin from *Rubus fruticosus* fruit ethyl acetate extract [30]. Sambandam et al. (2016) extracted and isolated quercetin from *Trigonella foenum-graecum* leaf ethanol extract using maceration and column chromatography, respectively [31]. Also, quercetin has been extracted from *Psidium guajava* leaf using maceration

Figure 3.
Some natural sources of quercetin.

by sonication in methanol: water (85:15) as the eluting solvent [32]. From the methanolic extract of *Ginkgo biloba*, quercetin has been extracted using solid-phase extraction (SPE) [33]. Pan and Lü reported the extraction and isolation of quercetin from *Gynostemma pentaphyllum* methanol extract using heat-reflux and high-speed countercurrent chromatography (HSCCC) methods, respectively [34]. Zhang et al. (2014) reported the extraction and isolation of quercetin from *Zanthoxylum bungeanum* 95% ethanol extract using maceration and column chromatography, respectively [35]. Additionally, quercetin has been extracted from *Dendrobium officinale* using the ultrasonic-assisted extraction (UAE) technique [36]. Pilařová et al. reported the extraction of quercetin from *A. praecox* extract using the carbon dioxide expanded liquid method [37].

3. Section 2: health benefits of quercetin

Quercetin has received great attention from scientists and researchers in the field of medicinal chemistry due to its vital therapeutic benefits and pharmacological properties. Many previous reports indicated the powerful ability of quercetin to

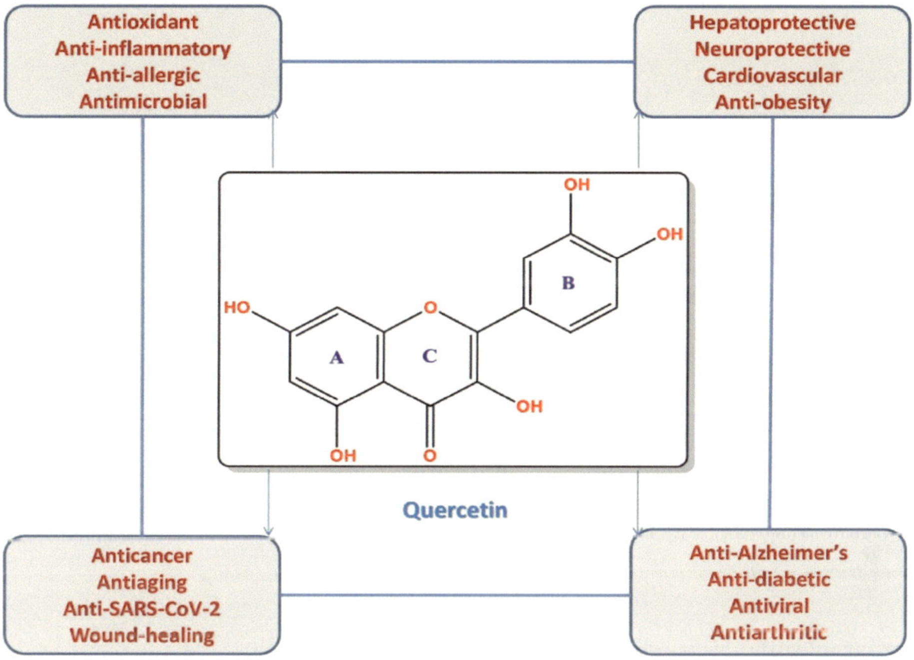

Figure 4.
Some pharmacological activities of quercetin.

treat many health disorders such as inflammations, oxidative stress, diabetes, viral, and microbial infections [38]. It has shown several biological properties such as antioxidant, antiarthritic, anti-inflammatory, anti-SARS-CoV-2, antiaging, antimicrobial, anti-obesity, cardiovascular, hepatoprotective, neuroprotective, anti-allergic, wound-healing, antiviral, anti-Alzheimer's, anti-diabetic, and anticancer activities (**Figure 4**) [39, 40]. Herein, some pharmacological activities of quercetin are mentioned in depth.

3.1 Antioxidant properties and mechanisms

Quercetin is known for its powerful antioxidant activity due to its ability to remove free radicals and protect cells from the accumulation of such serious species inside the body. It also has the ability to diminish the harmful effects resulting from the phenomenon of oxidative stress as a result of continuous exposure to some chemicals and pollutants. In addition to its potent free radical scavenging properties, quercetin has the ability to chelate with some metal ions such as Cu^{2+}, Mn^{2+}, and Fe^{2+} which increases its bioavailability [11, 41, 42].

Noteworthy, the structural effect of quercetin has a clear impact on its free radical scavenging activity. From the structure-activity relationship point of view, quercetin has the optimal structural requirements to achieve strong free radical scavenging activity including the presence of ortho-dihydroxy groups on the B-ring (catechol groups), unsaturated double bond between C-2 & C-3 and a carbonyl group (keto group) at C-4 on the C-ring, hydroxyl groups on the A- and C-rings at C-5 and C-3 positions, respectively, and 4-oxo group on the C-ring (**Figure 5**). These characteristic structural features led to stabilization of the aroxyl radical (**Ar-O•**) in the B-ring

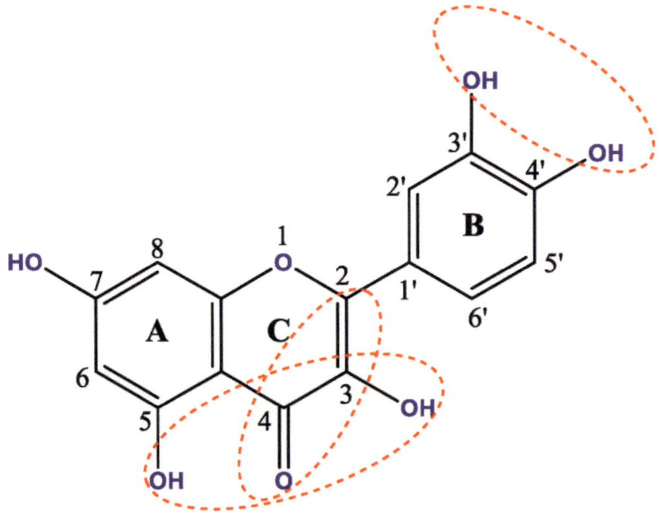

Figure 5.
Optimal structural criteria of quercetin for its effective free radical scavenging activity.

via intramolecular H-bonding. Moreover, the planarity of the molecule plays a vital role in the powerful activity of quercetin as an antioxidant and free radical scavenger natural agent [43–45].

3.2 Quercetin in disease prevention and management

Numerous previous reports have demonstrated the effective role of quercetin in disease prevention and management, including cardiovascular, cancer, and diabetes diseases [5, 38, 39]. Therefore, the higher daily quercetin intake is correlated to a minimal threat of several diseases.

3.2.1 Cardiovascular health

Previous reports revealed that the intake of quercetin-rich foods led to the prevention of cardiovascular diseases. In the same context, some studies conducted on experimental animals indicate a decrease in blood pressure after receiving nutritional supplements containing quercetin, confirming the effect of quercetin as a vasodilator agent [46]. Taken together, quercetin has the ability to decrease low-density lipoprotein as well as cholesterol oxidation and to inhibit endothelial dysfunction in cardiovascular diseases [47]. Herein, **Figure 6** summarized the cardioprotective effects of quercetin and its molecular mechanisms involved in cardioprotective management.

Numerous studies showed the cardioprotective role of quercetin using experimental animal and human models. Rasheed et al. reported the cardioprotective activity of quercetin and its role in the improvement of myocardial injury resulting from high-fat diet intake in adult male albino rat model [48]. Furthermore, quercetin has the ability to relieve endothelial dysfunction in age-related cardiovascular cases based on its anti-atherosclerotic and anti-hypertensive activities [49]. In another study, quercetin supplementation at a dose of (250 mg/day) for 2 months led to the improvement of oxidative stress, blood pressure, left ventricular function, and aerobic power in men with hypertension and coronary artery diseases [50]. Additionally,

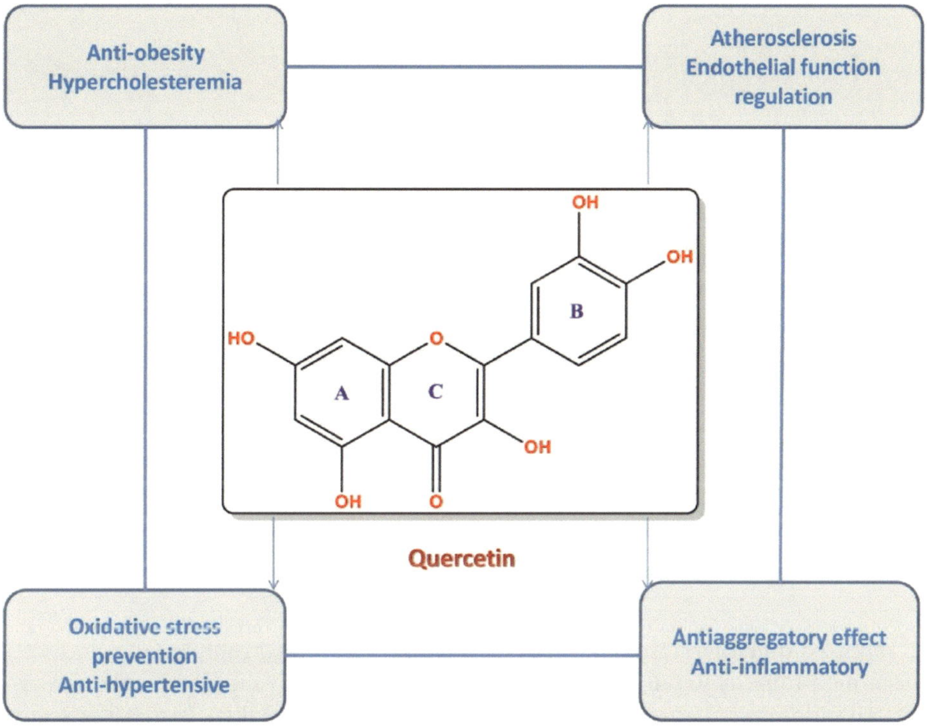

Figure 6.
The cardioprotective effects of quercetin and its molecular mechanisms.

quercetin could invert posttraumatic cardiac dysfunction via decreasing cardiomyocyte apoptosis across the repression of TNF-α rises, reactive species overproduction, and Ca^{2+} overburden in cardiomyocytes [51].

3.2.2 Anti-cancer properties

Quercetin is known for its chemopreventive and anti-cancer properties due to its unique chemical structure, which increases its ability to scavenge free radicals and enhance its antioxidant activity. Chemotherapy is one of the approaches used to combat cancerous diseases, in addition to surgery, radiotherapy, immunotherapy, targeted therapy, and hormone therapy. Noteworthy, quercetin has multiple mechanisms of action in treating cancer, such as cell cycle arrest, autophagy, potentiates apoptosis, drug resistance, and prevents angiogenesis and metastasis [52–54]. Quercetin has a long history in treating various types of cancer such as breast, colorectal, ovarian, blood, lung, oral, esophageal, bladder, gastric, hepatic and prostate cancers [55, 56]. The targeted cancers and the possible mechanisms of action are summarized in **Figure 7**.

3.2.3 Diabetes and metabolic syndrome

Previous reports have shown that the daily intake of quercetin reduces most signs of metabolic syndrome, including high blood pressure, high blood sugar, insulin resistance, abdominal obesity, and abnormal cholesterol. Additionally, quercetin has the superior ability to manage type 2 diabetes and its associated dependencies [57–59]. Desai

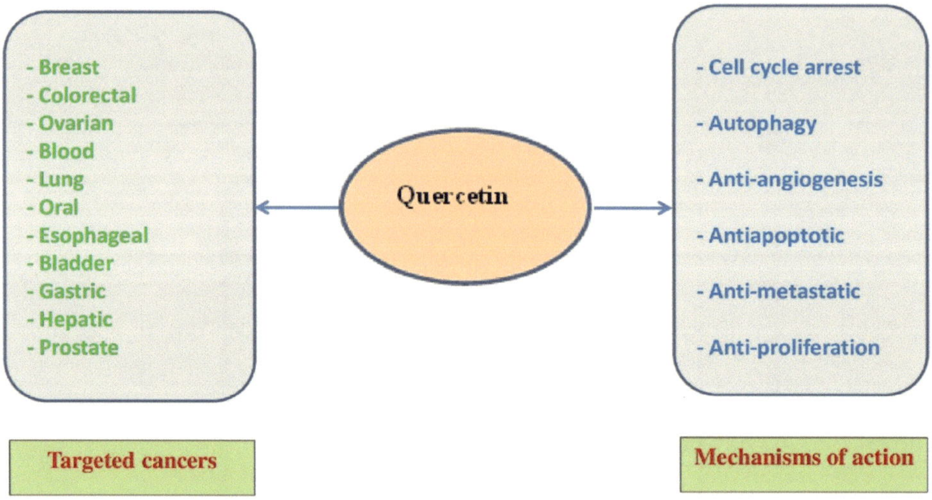

Figure 7.
The targeted cancers and the possible mechanisms of action.

et al. reported that quercetin treatment for 8 weeks alleviates streptozotocin-induced diabetic nephropathy in rats by controlling hyperglycemia, dyslipidemia, and down-regulated inflammatory activators NFκB, IL-6, and Caspase-3 [60]. In another study, quercetin treatment (70 mg/kg) alleviates diabetic encephalopathy via SIRT1/ER stress pathway in db/db mice [61]. In the same context, quercetin isolated from *Edgeworthia gardneri* flowers led to amelioration of type 2 diabetes mellitus by inducing insulin secretion [62]. Also, quercetin at a dose of (0.5 g/kg) led to the reduction of anxiety and decreasing hyperglycemia-related injury in induced diabetic's rats [63]. Moreover, the administration of quercetin at a dose of (15 mg/kg, intraperitoneally) led to the reduction of fasting blood sugar and malondialdehyde levels in rats with Streptozocin-induced diabetes [64]. Additionally, Rahmani and his co-workers reported the efficacy of quercetin (50 mg/kg b.w.) to reduce diabetic complications, renal tissue damage, and renal oxidative stress in Streptozotocin-induced diabetic rats [65].

3.3 Quercetin in mental health and neuroprotection

The neuroprotective properties of quercetin are well-known versus neurotoxins, neuronal injury, and neurodegenerative diseases [66]. Recently, several studies have demonstrated the efficacy of quercetin as a possible therapeutic agent in central nervous system syndromes such anti-depressive, anti-anxiolytic, anti-neuroinflammatory, and anti-Huntington [67]. Also, quercetin has received great attention from researchers in the treatment of neurological and mental disorders, including ischemia, cognitive impairment, traumatic injury, Alzheimer's, Parkinson's, and Huntington's complaints [68, 69]. Mehany et al. reported on the efficacy of quercetin in mending brain injury and cerebral alterations arising from high altitude, low pressure, and low oxygen through increasing antioxidant enzymes such as SOD, CAT, and GPx [70]. Regarding Alzheimer's diseases, quercetin plays a vital neuroprotective role in this issue by modulating learning, memory, and cognitive tasks [69]. Quercetin also has a high ability to inhibit neuroinflammatory routes via downregulating pro-inflammatory cytokines, like NF-kB and iNOS [66].

4. Section 3: toxicity and safety profile

Quercetin is highly regarded in the scientific community for its robust antioxidant properties, which play a crucial role in mitigating oxidative stress linked to chronic diseases such as cardiovascular disorders, diabetes, and cancer [71]. As a scavenger of free radicals, quercetin effectively combats cellular damage and leverages its anti-inflammatory capabilities to mitigate chronic inflammation, often associated with various pathologies [72].

Additionally, quercetin's anti-carcinogenic traits demonstrate its potential as both a preventative and therapeutic agent in cancer care: inhibiting cancer cell proliferation, preventing their invasion, and inducing apoptosis [73]. Yet, despite these benefits, the safety and toxicity of quercetin warrant careful consideration. While quercetin from dietary sources is safe, encapsulating the natural balance, supplementation, especially in high doses, introduces risks, including potential kidney toxicity [74, 75].

Quercetin's interactions with bodily enzymes and drug transporters also highlight its profound biological impact but raise concerns about its ability to alter pharmaceutical metabolism, necessitating cautious use, particularly among those on medication [76].

The extensive research into quercetin's bioactive properties, including its antioxidant, anti-inflammatory, and antiviral effects, underscores its potential as a beneficial dietary supplement. However, the principle that "the dose makes the poison" is especially relevant when assessing quercetin's toxicity at high supplementation levels, which could pose risks such as cytotoxicity and potential mutagenicity under certain conditions [77–81]. This complex profile calls for a balanced approach to using quercetin, emphasizing the importance of moderation and professional guidance to harness its health benefits while avoiding adverse effects.

4.1 Toxicological insights from animal models and human clinical studies

Animal studies have been instrumental in uncovering the potential toxic effects of quercetin. Rodent models, for instance, have shown that while moderate doses of quercetin can be beneficial, excessive intake can lead to kidney and liver toxicity, highlighting the organ-specific vulnerability to quercetin's cytotoxic effects [82]. These studies serve as a cautionary tale, suggesting that the extrapolation of quercetin's benefits to humans requires a nuanced understanding of dose-dependent toxicity [83]. Human studies on quercetin supplementation provide valuable insights into its toxicological profile. While most clinical trials report no significant adverse effects from quercetin at moderate doses, the variability in individual metabolism and the potential for drug-nutrient interactions warrant consideration. The impact of quercetin on enzyme systems involved in drug metabolism, particularly the cytochrome P450 enzymes, poses a risk for altering the pharmacokinetics of concurrently administered medications, necessitating a cautious approach to supplementation [84].

4.2 Safe dosage and potential side effects

Quercetin, a flavonoid present in many fruits, vegetables, and grains, is celebrated for its potent antioxidant, anti-inflammatory, and anti-carcinogenic properties [73]. However, while the benefits of quercetin are substantial, careful consideration of dosage and potential side effects is crucial to maximize its therapeutic value and minimize risks [85].

Quercetin's ability to scavenge free radicals, reduce inflammation, and regulate cell cycles contributes to its potential to mitigate chronic diseases and protecting against some cancers. Nevertheless, the principle that "the dose makes the poison" is particularly applicable here, emphasizing the importance of determining a safe dosage [86].

Research indicates that quercetin consumed through the diet is safe and contributes to lower chronic disease risks. However, supplementation, typically ranging from 500 to 1000 mg per day, exceeds dietary levels and requires careful safety evaluation [87, 88]. Health experts and regulatory bodies consider short-term supplementation of up to 1000 mg per day generally safe for most adults, but the safety of long-term use remains uncertain [89].

Potential side effects of quercetin, even if mild and transient, include headaches and tingling sensations. More seriously, high doses could impact kidney function, highlighted by animal studies suggesting possible toxicity [90, 91]. Additionally, quercetin can interact with enzymes and drug transporters, potentially altering the effectiveness or safety of other medications. This makes it crucial for those on medication to consult healthcare providers before starting quercetin supplements [92].

4.3 Interactions with pharmaceuticals and nutrients

The interaction of dietary supplements like quercetin with pharmaceuticals is crucial, particularly given quercetin's antioxidant, anti-inflammatory, and anti-carcinogenic properties and its rising popularity as a supplement [93, 94]. This flavonoid can significantly affect the metabolism of drugs, primarily through its influence on cytochrome P450 (CYP) enzymes like CYP3A4, which metabolizes many common medications such as statins and certain antidepressants. Inhibiting CYP3A4, quercetin could increase these drugs' plasma concentrations, raising potential toxicity risks [95–97]. Quercetin also affects drug transport mechanisms, notably P-glycoprotein (P-gp), which controls the efflux of drugs from cells. By inhibiting P-gp, quercetin might alter the efficacy and safety of drugs that are P-gp substrates, like some chemotherapeutic agents and immunosuppressants [98, 99]. Beyond pharmaceuticals, quercetin's interactions with nutrients are significant. It can enhance the body's antioxidative defense by synergizing with vitamins C and E, beneficial under oxidative stress conditions [100]. However, its chelating properties may affect the absorption of essential minerals like iron and zinc, which could be a concern for those with specific dietary limitations or marginal mineral status [101, 102].

Given these dynamics, integrating quercetin into therapeutic regimens demands careful consideration of both its benefits and potential risks, requiring healthcare professionals to assess individual factors and provide personalized advice to ensure safe and effective use.

5. Section 4: cutting-edge advances in quercetin research

Quercetin research within nanotechnology is promising due to its potential to enhance bioavailability and targeted delivery, overcoming traditional limitations like poor solubility and rapid metabolism [103]. Nanotechnology-based systems aim to optimize quercetin's therapeutic use by reducing dosage and minimizing side effects [104]. Additionally, quercetin is being investigated for its synergistic effects with other treatments, particularly in cancer therapy, to improve outcomes and reduce adverse effects by modulating key biological pathways [105, 106].

Research also extends to neurodegenerative diseases, exploring quercetin's antioxidative and anti-inflammatory properties for potential treatments for Alzheimer's and Parkinson's diseases [107]. In sports science, studies focus on quercetin's role in reducing inflammation and oxidative stress, potentially enhancing athletic performance and recovery [108]. Lastly, quercetin's impact on aging and longevity is being explored, particularly its effects on cellular mechanisms like telomerase activity and sirtuins, suggesting its use in promoting healthy aging and extending lifespan [109, 110].

5.1 Nanotechnology and quercetin delivery systems

Nanotechnology, a pluridisciplinary scientific undertaking, involves formation and utilization of materials, systems, and devices on the nanometer scale and is presently undergoing explosive development on numerous fronts [111]. Particularly in drug delivery, nanotechnology is expected to spark innovation and play a critical role in several biomedical applications [112]. Pharmaceutical industries have become progressively concerned in nanotechnological developments due to their wide benefits, for instance, targeted drug delivery, modified release systems, and the aptitude to develop novel formulations that were previously not possible [113].

Phytopharmacological and phytochemical disciplines have already established the biological abilities and composition of several herbal products [114]. The majority of the biologically active components of extracts, such as tannins, terpenoids, and flavonoids, are highly water insoluble, have large molecular sizes, and exhibit poor absorption, resulting in loss of bioavailability and efficacy [115]. Several studies have shown that herbal remedies have good activity in assays in vitro, which are not reproducible in investigations *in vivo* [116, 117]. Quercetin is a promising candidate for the management of various diseases. However, because of its hydrophobicity, this flavanol has a relatively low bioavailability, which must be increased to fully explore its potential [118]. Nanotechnology is an attractive option to overcome this limitation since it can enhance quercetin bioavailability at the anticipated site and, subsequently, improve therapeutic capabilities [119]. Numerous studies focus a great deal of interest on the creation of nanotechnological strategies that attempt to determine the most effective way to encapsulate and deliver quercetin for various uses (**Figure 8**).

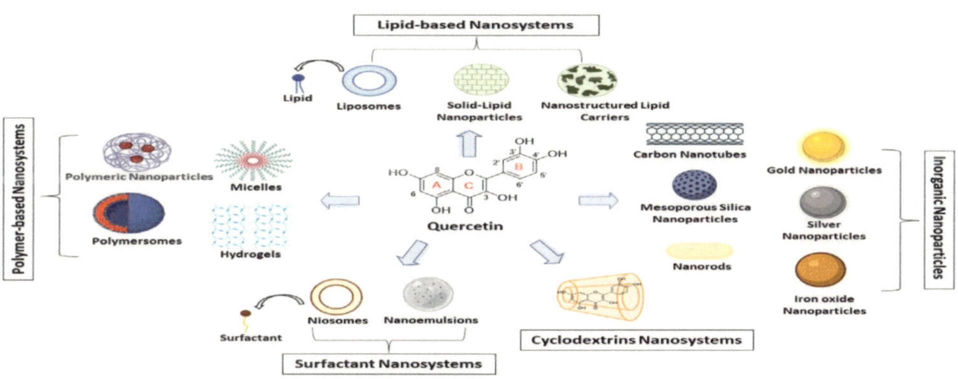

Figure 8.
Illustration of quercetin-delivery nanosystems [111].

5.2 Polymers

Polymers are striking materials for the pharmaceutical industry. They result from linking many high MW monomers, forming a large chain structure [120]. Nucleic acids, pharmaceutical agents, and other bioactive substances can be transported via polymer nanoparticulate platforms, which are broadly acknowledged as novel delivery methods [121]. Micelles are colloidal nanoparticles with a nanoscale of less than 100 nm and a distinct core-shell structure, formed of a hydrophobic core and a hydrophilic outer layer [122]. Pluronic F88 and P123 were used to create quercetin-incorporated micelles. The obtained formulation by Patel et al. offered an improved bioavailability of quercetin, a slower and controlled release pattern, and better ability to hinder the growth of MCF-7 cancer cells compared to pure quercetin [123]. Qi et al. proposed a highly promising quercetin nanocarrier through the formation of polymeric micelles using Soluplus, which is an amphiphilic graft copolymer. Results obtained from in-vivo studies displayed upgraded ability of quercetin-loaded micelles to reduce tumor size and decrease adverse effects. Moreover, Soluplus–Quercetin micelles possess a multi-faceted suppressing effect on angiogenesis, making them promising systems for the targeted delivery of quercetin in cancer management [124].

Nanoparticles (NPs) can be made of a variety of biomaterials, such as polymers, lipids, etc. Polymeric nanoparticles have widely attracted the attention of the scientific community [125]. Polymeric NPs are distinguished from other particles because of their size, ranging from 10 nm to 10 μm, as well as their distinctive physicochemical characteristics [126]. Numerous bioactive substances can be accommodated into the inner part or on the outer surface of NPs [127]. pH-sensitive NPs based on Eudragit® S100 polymer were prepared by Sunoqrot et al. for the targeted delivery of quercetin to colon cancer. The results of the cytotoxic profile of quercetin incorporated into NPs on CT26 murine colon cancer cells showed a higher antitumor potency compared to the free drug [111]. Also, Huang et al. developed an impressive delivery platform for quercetin utilizing polymeric NPs comprising PEG conjugated with polyethyleneimine (PEI). The results acknowledged that quercetin polymeric PEG–PEI NPs was able to reduce inflammation and oxidative stress, and limit the renal degradation that happened by acute kidney injury [128].

5.3 Hydrogels

Hydrogels are another class of polymeric systems that may be employed as quercetin delivery systems [129]. Hydrogels are categorized into natural, synthetic, as well as hybrids based on their source. Their three-dimensional configuration is attained by the cross-linking of polymers, which allows absorption of large volumes of aqueous solutions into their polymeric network [130]. Intriguingly, Mok et al. presented a quercetin delivery system made of a methoxy poly ethylene glycol-l-poly alanine polymer to reduce pain and slow the progression of osteoarthritis. The outcomes of in-vitro release studies established the extended release of quercetin for about 1 month and the lessening of cartilage degradation attributed to osteoarthritis. The reduction was detected either with the free quercetin hydrogel or with the quercetin-loaded one; however, the presence of quercetin in the formula was found to offer further protection against the disease [131]. A polymersome is a complex configuration with a distinctive "pseudo-spherical" shell created by the self-assembly behavior of amphiphilic block copolymers. Polymersome structure is able to accommodate hydrophilic substances into the aqueous interior and to interact with the lipophilic substances on the exterior membranes [121]. Gomes et al. developed a research focused on the encapsulation of polyphenols, such as quercetin into polymersomes. Subsequently, the research results confirmed the system's colloidal stability as well as its aptitude to be used as drug delivery system for polyphenols [132].

5.4 Lipid-based nanoparticles

Lipid-based nanoparticles (NPs), including liposomes and solid lipid nanoparticles (SLN), are increasingly used for drug delivery due to their high encapsulation efficiency and ease of surface modification [133, 134]. Liposomes, mimicking cellular membranes, consist of hydrophilic and hydrophobic phospholipids that spontaneously form bilayers in aqueous solutions, influencing their size, charge, and rigidity [135, 136]. These liposomes are particularly effective in delivering low molecular weight drugs and genes [134–136] and are biocompatible and minimally immunogenic [137–140]. Studies have demonstrated liposomes' ability to encapsulate and enhance the delivery of quercetin, a therapeutic agent, across various applications. For example, Liu et al. developed cholesterol-based liposomes for protecting against UVB-induced skin damage [141], while Shaji et al. and Yuan et al. used liposomal quercetin for liver protection and tumor suppression respectively [142, 143]. Additionally, Wong et al. utilized liposomes to deliver quercetin and vincristine to treat breast tumors [144].

For brain-related therapies, liposomal quercetin has shown potential in reducing anxiety and enhancing cognition in animal models [145, 146]. Beyond liposomes, SLN and nanostructured lipid carriers (NLC) offer benefits like physical stability and controlled drug release, though they may face issues like lipid recrystallization [147–150]. Various formulations have been tested to increase quercetin's bioavailability and efficacy, such as Li et al.'s SLN which enhanced gastrointestinal absorption [151, 152], and Dhawan et al.'s neuroprotective SLN that improved memory retention in a dementia model [153, 154].

5.5 Surfactant-based nanoparticles

Niosomes, nanovesicles made from non-ionic surfactants and lipids such as cholesterol, self-assemble into bilayers in water, encapsulating drugs effectively due to their amphiphilic nature [155–160]. Recent studies highlight their use in delivering the antioxidant quercetin, enhancing its stability and delivery efficiency. Innovations include hyaluronic acid niosomes for anti-inflammatory effects in rats [161] and sugar esters-based niosomes for liver protection in rat models [159]. Additionally, quercetin combined with captopril in niosomes showed prolonged release and significant health benefits [162]. Nanoemulsions, colloidal dispersions of nanosized particles, demonstrate prolonged drug release and targeted cytotoxic actions in various formulations [163, 164]. A quercetin-based palm oil nanoemulsion showed potential against hepatic cancer cells [165], and other nanoemulsions improved quercetin bioavailability in diabetic rat models [166]. Inorganic nanoparticles, especially magnetic ones, are promising for targeted quercetin delivery to tumors, utilizing external magnetic fields [167–169]. Notably, Fe_3O_4 nanoparticles and SPIONs optimized with functional coatings have shown effective targeting and controlled release [170, 171]. Similarly, mesoporous silica nanoparticles (MSN), known for their stability and high drug load capacity, have shown potential in delivering quercetin for cancer therapy and skin penetration enhancement [172–175]. Cyclodextrins (CDs), cyclic oligosaccharides, offer limited but effective quercetin encapsulation, potentially increasing its antioxidant effects. Specific types of CDs have shown varying degrees of effectiveness in cancer and antioxidant applications [176–180].

5.6 Challenges in quercetin research and application

One of the primary challenges in quercetin's application lies in its bioavailability and solubility. Quercetin has poor water solubility, which significantly hampers its absorption and bioavailability when ingested. This limitation restricts the amount of quercetin that enters the circulation and reaches the target tissues, thereby diminishing its therapeutic effectiveness [181, 182]. Researchers are exploring various strategies, including nanoparticle delivery systems and complexation with cyclodextrins, to enhance the solubility and bioavailability of quercetin. While these approaches show promise, they also introduce additional complexity and cost to the development of quercetin-based therapies [183]. Another challenge is the lack of standardized dosing guidelines for quercetin supplementation. The optimal dose of quercetin can vary significantly depending on the condition being treated, the form of quercetin used (e.g., quercetin dihydrate vs. aglycone), and individual factors such as age, gender, and metabolic health [184]. Establishing comprehensive dosing guidelines requires extensive clinical trials across diverse populations, a process that is both time-consuming and costly. Without standardized dosing, the efficacy and safety of quercetin supplementation remain variable, complicating its clinical application [185].

Quercetin's interactions with pharmaceuticals present a significant challenge in its application. As an inhibitor of several cytochrome P450 enzymes and a modulator of drug transporters like P-glycoprotein, quercetin can affect the metabolism and efficacy of various medications. This interaction poses risks of adverse effects or reduced therapeutic effectiveness of drugs, particularly in polypharmacy scenarios common in chronic disease management. Understanding and mitigating these interactions requires detailed pharmacokinetic studies and personalized medicine approaches, adding layers of complexity to quercetin's clinical use [95].

The long-term safety of quercetin supplementation is yet another area that necessitates further investigation. While quercetin is generally considered safe when consumed in amounts typically found in a diet rich in fruits and vegetables, the safety of higher doses used in supplements over extended periods is less clear. Potential concerns include kidney toxicity and interactions with hormone-sensitive conditions. Longitudinal studies are crucial to assess the long-term safety of quercetin and provide clear guidance for its use [186].

5.7 Future directions in quercetin studies

A pivotal area of future research lies in addressing quercetin's limited bioavailability. Innovative strategies such as nanotechnology-based delivery systems, liposomal encapsulation, and complexation with other molecules promise to enhance the solubility, stability, and tissue targeting of quercetin. Further development and optimization of these delivery methods are crucial to maximize quercetin's therapeutic efficacy. Investigations into the pharmacokinetics of these novel formulations will provide valuable insights into optimizing dosage and administration routes, paving the way for more effective quercetin-based therapies [187–189].

While preclinical studies have provided substantial evidence of quercetin's beneficial effects, there is a pressing need for comprehensive, large-scale clinical trials to validate these findings in human populations. Future research should focus on diverse demographic groups, encompassing various ages, genders, and ethnic backgrounds to ensure the generalizability of results. Clinical trials designed to investigate specific health outcomes, dosage optimizations, and long-term safety will be instrumental in establishing quercetin's role in preventing and treating diseases, as well as in dietary recommendations and supplement formulations [85, 190, 191]. Understanding the precise mechanisms through which quercetin exerts its effects remains a significant challenge. Future studies should delve deeper into the molecular and cellular pathways modulated by quercetin, employing cutting-edge techniques such as transcriptomics, proteomics, and metabolomics. Research into the epigenetic modifications induced by quercetin could uncover novel mechanisms of action and potential therapeutic targets. This mechanistic insight will not only advance our understanding of quercetin's biological activities but also facilitate the identification of synergistic interactions with other compounds and drugs [192–195]. Given the rising prevalence of chronic diseases worldwide, quercetin's potential role in management and prevention strategies warrants further exploration. Future research should aim to elucidate the efficacy of quercetin supplementation in chronic disease conditions such as cardiovascular diseases, diabetes, neurodegenerative disorders, and various forms of cancer. Studies evaluating the impact of quercetin on disease progression, quality of life, and survival outcomes will be particularly valuable. Additionally, exploring the preventive aspects of quercetin in at-risk populations could lead to novel strategies for disease prevention and health promotion [196, 197]. The field of personalized medicine offers exciting possibilities for tailoring quercetin-based interventions to individual genetic profiles and health conditions. Future directions in quercetin research should include investigations into the interplay between genetic variations, quercetin metabolism, and therapeutic outcomes. This personalized approach could optimize the benefits of quercetin supplementation, minimize potential adverse effects, and identify populations that may derive the most significant benefits from quercetin interventions [105, 198].

Lastly, research into the environmental and agricultural factors affecting quercetin content in foods could enhance our understanding of how to maximize dietary quercetin intake through food sources. Studies on the effects of soil quality, climate change, and farming practices on the quercetin concentrations in crops could inform agricultural guidelines and dietary recommendations, promoting optimal health through natural dietary sources of quercetin [199].

5.8 Ethical and regulatory considerations

As quercetin—a naturally occurring flavonoid known for its antioxidant and anti-inflammatory properties—gains popularity, it is crucial to address the ethical and regulatory aspects that shape its research and usage [200]. This focus is driven by its increasing application in health-related products and the need for rigorous oversight to ensure public safety and scientific integrity [201].

Ethically conducting quercetin research involves safeguarding participant welfare and rights in clinical trials. This includes adhering to ethical standards like informed consent, privacy, and equitable participant selection. Itis essential to communicate the risks and benefits transparently and to ensure that research findings are reported accurately, without overstating quercetin's effects or underreporting its potential risks [202, 203].

Regulatory oversight is vital for turning quercetin research into safe and effective treatments. Unlike pharmaceuticals, dietary supplements containing quercetin might not undergo stringent testing, leading to variability in product quality. Regulatory bodies need to enforce strict standards for manufacturing, labeling and claims substantiation to guarantee that health claims are scientifically backed and conveyed accurately [204–206]. Additionally, the marketing of quercetin supplements should responsibly reflect the scientific evidence to avoid misleading consumers. Regulators and ethical bodies must ensure that promotional materials are truthful, helping consumers make informed decisions and preventing undue reliance on supplements for serious health issues. Strict guidelines and enforcement are essential to balance promoting quercetin's health benefits with protecting public and consumer interests [207, 208].

6. Conclusion

Quercetin, with its robust presence in the plant kingdom, serves as a cornerstone of the flavonoid family, offering a diverse array of health benefits owing to its potent antioxidant, anti-inflammatory, and anti-carcinogenic properties. Its molecular architecture enables a multitude of pharmacological activities, making it a beacon of hope against chronic diseases such as cardiovascular ailments, diabetes, and various forms of cancer. The historical and contemporary interest in quercetin underscores its significance not only as a nutritional component but also as a potential therapeutic agent, drawing attention to its integration into pharmaceutical formulations, cosmetics, and food production. However, the leap from dietary component to therapeutic agent is bridled with challenges, notably concerning bioavailability, toxicity, and interaction with pharmaceuticals. Quercetin's poor solubility and rapid metabolism have spurred research into innovative delivery systems, including nanotechnology-based solutions, which promise to enhance its bioavailability and therapeutic efficacy. Yet, these advances bring forth regulatory and ethical considerations, especially regarding dosage standardization and long-term safety. The potential adverse effects and toxicity associated with high doses of quercetin or its interaction with drugs underscore the necessity of a prudent approach to supplementation. Ethical and regulatory frameworks must evolve in tandem with scientific advancements to ensure quercetin's safe and effective application. The exploration of quercetin's interactions with pharmaceuticals and nutrients further highlights the complexity of its integration into therapeutic regimens, necessitating personalized medicine approaches to optimize its benefits while minimizing risks. Cutting-edge research into nanotechnology and targeted delivery systems opens new vistas for quercetin's application, promising to overcome the barriers of bioavailability and offering new strategies for disease prevention and treatment. These innovations, coupled with a deeper understanding of quercetin's mechanisms of action and its impact on chronic diseases, portend a future where quercetin could play a pivotal role in enhancing human health.

In conclusion, quercetin stands at the crossroads of tradition and innovation, embodying the promise of natural compounds in advancing human health. As research continues to unravel its complexities, a balanced approach—rooted in scientific rigor, ethical considerations, and regulatory oversight—is essential. By navigating the challenges and harnessing the advances in quercetin research, the scientific community can unlock its full potential, paving the way for novel therapeutic strategies that capitalize on the synergies between natural compounds and cutting-edge technology.

Acknowledgement

The authors acknowledge the use of QuillBot for language polishing of the manuscript.

Author details

Mosad A. Ghareeb[1*], Abdallah Z. Zayan[2], Falah H. Shari[3] and Ahmed M. Sayed[4]

1 Medicinal Chemistry Department, Theodor Bilharz Research Institute Kornaish El Nile, Warrak El-Hadar, Imbaba, Giza, Egypt

2 Department of Pharmaceutics, Collage of Pharmacy, Almaaqal University, Basrah, Iraq

3 Collage of Pharmacy, Almaaqal University, Basrah, Iraq

4 Department of Pharmacognosy, Collage of Pharmacy, Almaaqal University, Basrah, Iraq

*Address all correspondence to: m.ghareeb@tbri.gov.eg

References

[1] Horwitz RJ. The allergic patient. In: Integrative Medicine. 4th ed. Amsterdam, Netherlands: Elsevier; 2018. pp. 300-309.e2. DOI: 10.1016/B978-0-323-35868-2.00030-X

[2] Jaffe R, Mani J. Chapter 29-polyphenolics evoke healing responses: Clinical evidence and role of predictive biomarkers. In: Watson RR, Preedy VR, Zibadi S, editors. Polyphenols: Mechanisms of Action in Human Health and Disease. 2nd ed. London, United Kingdom: Academic Press; 2018. pp. 403-413. DOI: 10.1016/B978-0-12-813006-3.00029-5

[3] Anand David AV, Arulmoli R, Parasuraman S. Overviews of biological importance of quercetin: A bioactive flavonoid. Pharmacognosy Reviews. 2016;**10**(20):84-89. DOI: 10.4103/0973-7847.194044

[4] Shabir I, Kumar Pandey V, Shams R, Dar AH, Dash KK, Khan SA, et al. Promising bioactive properties of quercetin for potential food applications and health benefits: A review. Frontiers in Nutrition. 2022;**9**:999752. DOI: 10.3389/fnut.2022.999752

[5] Shebeko SK, Zupanets IA, Popov OS, Tarasenko OO, Shalamay AS. Chapter 27-effects of quercetin and its combinations on health. In: Watson RR, Preedy VR, Zibadi S, editors. Polyphenols: Mechanisms of Action in Human Health and Disease. 2nd ed. London, United Kingdom: Academic Press; 2018. p. 373, 394. DOI: 10.1016/B978-0-12-813006-3.00027-1

[6] Hurjui LL, Hartan RM, Hurjui IA, Tărniceriu CC, Hurjui I, Jipu R, et al. Quercetin in health and disease. Romanian Journal of Oral Rehabilitation. 2021;**13**(2):223-230

[7] Ay M, Charli A, Jin H, Anantharam V, Kanthasamy A, Kanthasamy AG. Quercetin. In: Nutraceuticals-Efficacy, Safety and Toxicity. 1st ed. Chapter 32. Amsterdam, Netherlands: Elsevier; 2016. pp. 447-452. DOI: 10.1016/B978-0-12-802147-7.00032-2

[8] Hertog MG, Kromhout D, Aravanis C, Blackburn H, Buzina R, Fidanza F, et al. Flavonoid intake and long-term risk of coronary heart disease and cancer in the seven countries study. Archives of Internal Medicine. 1995;**155**(4):381-386

[9] Dhanasekaran S, Jaganathan R. Augmented cell signaling by natural polyphenols and flavonoids: Insights into cancer cell remodeling and prevention. In: Atta-ur-Rahman, editor. Studies in Natural Products Chemistry. Amsterdam, Netherlands: Elsevier; 2018. pp. 213-244. DOI: 10.1016/B978-0-444-64056-7.00007-6

[10] Shen N, Wang T, Gan Q, Liu S, Wang L, Jin B. Plant flavonoids: Classification, distribution, biosynthesis, and antioxidant activity. Food Chemistry. 2022;**383**:132531. DOI: 10.1016/j.foodchem.2022.132531

[11] Molina MF, Sanchez-Reus I, Iglesias I, Benedi J. Quercetin, a flavonoid antioxidant, prevents and protects against ethanol-induced oxidative stress in mouse liver. Biological & Pharmaceutical Bulletin. 2003;**26**(10):1398-1402. DOI: 10.1248/bpb.26.1398

[12] Hrelia S, Angeloni C. Quercetin and its metabolites in heart health. In: Watson RR, Preedy VR, editors. Bioactive Food as Dietary Interventions for Cardiovascular Disease. London, United Kingdom: Academic Press;

2013. pp. 217-228. DOI: 10.1016/B978-0-12-396485-4.00011-6

[13] Williams CA, Grayer RJ. Anthocyanins and other flavonoids. Natural Product Reports. 2004;**21**(4):539-573. DOI: 10.1039/b311404j

[14] Materska M. Quercetin and its derivatives: Chemical structure and bioactivity-A review. Polish Journal of Food and Nutrition Sciences. 2008;**58**(4):407-413

[15] Hassan JK, Sharrad AK, Sheri FH. Effect of quercetin supplement on some bone mineralization biomarkers in diabetic type 2 patients. Advances in Pharmacology and Pharmacy. 2018;**6**:43-49. DOI: 10.13189/app.2018.060202

[16] Bansal M, Singh N, Pal S, Dev I, Ansari KM. Chemopreventive role of dietary phytochemicals in colorectal cancer. In: Fishbein JC, Heilman JM, editors. Advances in Molecular Toxicology. Amsterdam, Netherlands: Elsevier; 2018. pp. 69-121. DOI: 10.1016/B978-0-444-64199-1.00004-X

[17] Moodi Z, Bagherzade G, Peters J. Quercetin as a precursor for the synthesis of novel nanoscale Cu (II) complex as a catalyst for alcohol oxidation with high antibacterial activity. Bioinorganic Chemistry and Applications. 2021;**2021**:8818452. DOI: 10.1155/2021/8818452

[18] Shankar GM, Antony J, Anto RJ. Quercetin and tryptanthrin: Two broad spectrum anticancer agents for future chemotherapeutic interventions. In: Bathaie SZ, Tamanoi F, editors. The Enzymes. London, United Kingdom: Academic Press; 2015. pp. 43-72. DOI: 10.1016/bs.enz.2015.05.001

[19] Nabil-Adam A, Elnosary ME, Ashour ML, Abd El-Moneam NM, Shreadah MA. Flavonoids Biosynthesis in Plants as a Defense Mechanism: Role and Function Concerning Pharmacodynamics and Pharmacokinetic Properties. London, United Kingdom: IntechOpen; 2023. DOI: 10.5772/intechopen.108637

[20] Marıín L, Gutiérrez-del-Río I, Entrialgo-Cadierno R, Villar CJ, Lombó F. *De novo* biosynthesis of myricetin, kaempferol and quercetin in *Streptomyces albus* and *Streptomyces coelicolor*. PLoS One. 2018;**13**(11):e0207278. DOI: 10.1371/journal.pone.0207278

[21] Ulusoy HG, Sanlier N. A minireview of quercetin: From its metabolism to possible mechanisms of its biological activities. Critical Reviews in Food Science and Nutrition. 2020;**60**(19):3290-3303. DOI: 10.1080/10408398.2019.1683810

[22] Yi H, Peng H, Wu X, Xu X, Kuang T, Zhang J, et al. The therapeutic effects and mechanisms of quercetin on metabolic diseases: Pharmacological data and clinical evidence. Oxidative Medicine and Cellular Longevity. 2021;**2021**:6678662. DOI: 10.1155/2021/6678662

[23] Martino KG, Guyer D. Supercritical fluid extraction of quercetin from onion skins. Journal of Food Process Engineering. 2004;**27**(1):17-28. DOI: 10.1111/j.1745-4530.2004.tb00620.x

[24] Wach A, Pyrzyńska K, Biesaga M. Quercetin content in some food and herbal samples. Food Chemistry. 2007;**100**(2):699-704. DOI: 10.1016/j.foodchem.2005.10.028

[25] Jang M, Asnin L, Nile SH, Keum YS, Kim HY, Park SW. Ultrasound-assisted extraction of quercetin from onion solid wastes. International Journal of Food Science and Technology. 2013;**48**(2):246-252. DOI: 10.1111/j.1365-2621.2012.03180.x

[26] Kumar B, Smita K, Kumar B, Cumbal L, Rosero G. Microwave-assisted extraction and solid-phase separation of quercetin from solid onion (*Allium cepa* L.). Separation Science and Technology. 2014;**49**(16):2502-2509. DOI: 10.1080/01496395.2014.933982

[27] Thuy NM, Tuyen NTM, Cuong NP, Huyen LTN, Phuong NP, Nguyen LTT, et al. Identification and extraction method of quercetin from flesh and skin of shallot (*Allium ascalonicum*) cultivated in Soc Trang province, Vietnam. Food Research. 2020;**4**(2):358-365

[28] Horbowicz M. Method of quercetin extraction from dry scales of onion. Vegetable Crops Research Bulletin. 2002;**57**:119-124

[29] Pandiyan R, Ilango K. Isolation, characterization, and validation of RP-HPLC method for the quantification of quercetin in *Huberantha senjiana* leaf extract. Journal of Applied Pharmaceutical Science. 2020;**10**(05):110-118. DOI: 110.7324/JAPS.2020.10515

[30] Zahoor M, Shah AB, Naz S, Ullah R, Bari A, Mahmood HM. Isolation of quercetin from *Rubus fruticosus*, their concentration through NF/RO membranes, and recovery through carbon nanocomposite. A pilot plant study. BioMed Research International. 2020;**2020**:8216435. DOI: 10.1155/2020/8216435

[31] Sambandam B, Thiyagarajan D, Ayyaswamy A, Raman P. Extraction and isolation of flavonoid quercetin from the leaves of *Trigonella foenum-graecum* and their anti-oxidant activity. International Journal of Pharmacy and Pharmaceutical Sciences. 2016;**8**(6):120-124

[32] Batubara I, Suparto IH, Wulandari NS. The best extraction technique for kaempferol and quercetin isolation from guava leaves (*Psidium guajava*). IOP Conference Series: Earth and Environmental Science. 2017;**58**:012060. DOI: 10.1088/1755-1315/58/1/012060

[33] Wang X, Li G, Row KH. Extraction and determination of quercetin from *Ginkgo biloba* by DESs-based polymer monolithic cartridge. Journal of Chromatographic Science. 2017;**55**(8):866-871. DOI: 10.1093/chromsci/bmx037

[34] Pan C, Lü H. Preparative separation of quercetin, ombuin and kaempferide from *Gynostemma pentaphyllum* by high-speed countercurrent chromatography. Journal of Chromatographic Science. 2019;**57**(3):265-271. DOI: 10.1093/chromsci/bmy110

[35] Zhang Y, Wang D, Yang L, Zhou D, Zhang J. Purification and characterization of flavonoids from the leaves of *Zanthoxylum bungeanum* and correlation between their structure and antioxidant activity. PLoS One. 2014;**9**(8):e105725. DOI: 10.1371/journal.pone.0105725

[36] Zhu Y, Yu J, Jiao C, Tong J, Zhang L, Chang Y, et al. Optimization of quercetin extraction method in *Dendrobium officinale* by response surface methodology. Heliyon. 2019;**5**(9):e02374. DOI: 10.1016/j.heliyon.2019.e02374

[37] Pilařová V, Kuda L, Vlčková HK. Carbon dioxide expanded liquid: An effective solvent for the extraction of quercetin from South African medicinal plants. Plant Methods. 2022;**18**:87. DOI: 10.1186/s13007-022-00919-6

[38] Wang G, Wang Y, Yao L, Gu W, Zhao S, Shen Z, et al. Pharmacological activity of quercetin: An updated review. Evidence-Based Complementary and

Alternative Medicine. 2022;**2022**:3997190. DOI: 10.1155/2022/3997190

[39] Salehi B, Machin L, Monzote L, Sharifi-Rad J, Ezzat SM, Salem MA, et al. Therapeutic potential of quercetin: New insights and perspectives for human health. ACS Omega. 2020;**5**(20):11849-11872. DOI: 10.1021/acsomega.0c01818

[40] Asgharian P, Tazehkand AP, Soofiyani SR, Hosseini K, Martorell M, Tarhriz V, et al. Quercetin impact in pancreatic cancer: An overview on its therapeutic effects. Oxidative Medicine and Cellular Longevity. 2021;**2021**: 4393266. DOI: 10.1155/2021/4393266

[41] Paganga G, Al-Hashim H, Khodr H, Scott BC, Aruoma OI, Hider RC, et al. Mechanisms of antioxidant activities of quercetin and catechin. Redox Report: Communications in Free Radical Research. 1996;**2**(6):359-364. DOI: 10.1080/13510002.1996.11747075

[42] Murakami Y, Kawata A, Ito S, Katayama T, Fujisawa S. Radical-scavenging and anti-inflammatory activity of quercetin and related compounds and their combinations against RAW264.7 cells stimulated with Porphyromonas gingivalis fimbriae. Relationships between anti-inflammatory activity and quantum chemical parameters. *In vivo* (Athens, Greece). 2015;**29**(6):701-710

[43] Rice-Evans CA, Miller NJ, Paganga G. Structure-antioxidant activity relationships of flavonoids and phenolic acids. Free Radical Biology & Medicine. 1996;**20**(7):933-956. DOI: 10.1016/0891-5849(95)02227-9

[44] Rice-Evans C, Miller N, Paganga G. Antioxidant properties of phenolic compounds. Trends in Plant Science. 1997;**2**(4):152-159. DOI: 10.1016/S1360-1385(97)01018-2

[45] Platzer M, Kiese S, Tybussek T, Herfellner T, Schneider F, Schweiggert-Weisz U, et al. Radical scavenging mechanisms of phenolic compounds: A quantitative structure-property relationship (QSPR) study. Frontiers in Nutrition. 2022;**9**:882458. DOI: 10.3389/fnut.2022.882458

[46] Larson AJ, Symons JD, Jalili T. Therapeutic potential of quercetin to decrease blood pressure: Review of efficacy and mechanisms. Advances in Nutrition. 2012;**3**(1):39-46. DOI: 10.3945/an.111.001271

[47] Nazari-Khanamiri F, Ghasemnejad-Berenji M. Quercetin and heart health: From molecular pathways to clinical findings. Journal of Food Biochemistry. 2023;**2023**:8459095. DOI: 10.1155/2023/8459095

[48] Rasheed R, Othman M, Hussein U, Embaby A. The possible ameliorative influence of quercetin on cardiac muscle changes induced by high fat diet in adult male albino rats: Light and electron microscopic study. Egyptian Journal of Histology. 2022;**45**(3):937-948. DOI: 10.21608/ejh.2021.76868.1484

[49] Dagher O, Mury P, Thorin-Trescases N, Noly PE, Thorin E, Carrier M. Therapeutic potential of quercetin to alleviate endothelial dysfunction in age-related cardiovascular diseases. Frontiers in Cardiovascular Medicine. 2021;**8**:658400. DOI: 10.3389/fcvm.2021.658400

[50] Moonikh KO, Kashef M, Mahmoudi K, Salehpour M. Effects of quercetin supplementation on oxidative stress, blood pressure, aerobic power, concentric pathologic hypertrophy and cardiac function in men with hypertension and coronary artery disease after percutaneous coronary intervention: A randomized,

double-blind placebo-controlled trial. Nutrition and Food Sciences Research. 2020;7(2):21-28

[51] Jing Z, Wang Z, Li X, Li X, Cao T, Bi Y, et al. Protective effect of quercetin on posttraumatic cardiac injury. Scientific Reports. 2016;**6**:30812. DOI: 10.1038/srep30812

[52] Youn H, Jeong JC, Jeong YS, Kim EJ, Um SJ. Quercetin potentiates apoptosis by inhibiting nuclear factor-kappaB signaling in H460 lung cancer cells. Biological & Pharmaceutical Bulletin. 2013;**36**(6):944-951. DOI: 10.1248/bpb.b12-01004

[53] Tang SM, Deng XT, Zhou J, Li QP, Ge XX, Miao L. Pharmacological basis and new insights of quercetin action in respect to its anti-cancer effects. Biomedicine & Pharmacotherapy. 2020;**121**:109604. DOI: 10.1016/j.biopha.2019.109604

[54] Dogan M. Assessment of mechanism involved in the apoptotic and anti-cancer activity of quercetin and quercetin-loaded chitosan nanoparticles. Medical Oncology. 2022;**39**(11):176. DOI: 10.1007/s12032-022-01820-x

[55] Rauf A, Imran M, Khan IA, Ur-Rehman M, Gilani SA, Mehmood Z, et al. Anticancer potential of quercetin: A comprehensive review. Phytotherapy Research. 2018;**32**(11):2109-2130. DOI: 10.1002/ptr.6155

[56] Asgharian P, Tazekand AP, Hosseini K, Forouhandeh H, Ghasemnejad T, Ranjbar M, et al. Potential mechanisms of quercetin in cancer prevention: Focus on cellular and molecular targets. Cancer Cell International. 2022;**22**(1):257. DOI: 10.1186/s12935-022-02677-w

[57] Bilginoğlu A. Possible protective effect of quercetin against oxidative stress in liver from metabolic syndrome rats. Journal of Contemporary Medical Sciences. 2019;**9**(3):203-208

[58] Hosseini A, Razavi BM, Banach M, Hosseinzadeh H. Quercetin and metabolic syndrome: A review. Phytotherapy Research. 2021;**35**(10):5352-5364. DOI: 10.1002/ptr.7144

[59] Dhanya R. Quercetin for managing type 2 diabetes and its complications, an insight into multitarget therapy. Biomedicine & Pharmacotherapy. 2022;**146**:112560. DOI: 10.1016/j.biopha.2021.112560

[60] Desai A, Shah H, Patel A, Gandhi TR. Quercetin mitigates diabetic nephropathy in rats via Keap1/Nrf2/HO-1 signaling pathway. Journal of Natural Remedies. 2023;**23**(3):1009-1024. DOI: 10.18311/jnr/2023/33313

[61] Hu T, Shi JJ, Fang J, Wang Q, Chen YB, Zhang SJ. Quercetin ameliorates diabetic encephalopathy through SIRT1/ER stress pathway in db/db mice. Aging. 2020;**12**(8):7015-7029. DOI: 10.18632/aging.103059

[62] Zhuang M, Qiu H, Li P, Hu L, Wang Y, Rao L. Islet protection and amelioration of type 2 diabetes mellitus by treatment with quercetin from the flowers of *Edgeworthia gardneri*. Drug Design, Development and Therapy. 2018;**12**:955-966. DOI: 10.2147/DDDT.S153898

[63] Boudiaf F, Chouba I, Amri N, Tahraoui A. Anti-diabetic role of quercetin and cinnamon on neurobehavioral alterations and biochemical parameters of induced diabetics rats. Journal of Animal Behaviour and Biometeorology. 2020;**8**:190-195. DOI: 10.31893/jabb.20025

[64] Hemmati M, Mostafavi SE, Zarban A, Hoshyar R. Protective effects of quercetin on hyperglycemia and stress proteins expression in rats with Streptozocin-induced diabetes. Modern Care Journal. 2018;**15**(2):e64964. DOI: 10.5812/modernc.64964

[65] Rahmani AH, Alsahli MA, Khan AA, Almatroodi SA. Quercetin, a plant flavonol attenuates diabetic complications, renal tissue damage, renal oxidative stress and inflammation in Streptozotocin-induced diabetic rats. Metabolites. 2023;**13**:130. DOI: 10.3390/metabo13010130

[66] Costa LG, Garrick JM, Roquè PJ, Pellacani C. Mechanisms of neuroprotection by quercetin: Counteracting oxidative stress and more. Oxidative Medicine and Cellular Longevity. 2016;**2016**:2986796. DOI: 10.1155/2016/2986796

[67] Wróbel-Biedrawa D, Grabowska K, Galanty A, Sobolewska D, Podolak IA. Flavonoid on the brain: Quercetin as a potential therapeutic agent in central nervous system disorders. Life. 2022;**12**:591. DOI: 10.3390/life12040591

[68] Dhiman P, Malik N, Sobarzo-Sánchez E, Uriarte E, Khatkar A. Quercetin and related chromenone derivatives as monoamine oxidase inhibitors: Targeting neurological and mental disorders. Molecules. 2019;**24**(3):418. DOI: 10.3390/molecules24030418

[69] Khan H, Ullah H, Aschner M, Cheang WS, Akkol EK. Neuroprotective effects of quercetin in Alzheimer's disease. Biomolecules. 2019;**10**(1):59. DOI: 10.3390/biom10010059

[70] Mehany ABM, Belal A, Santali EY, Shaaban S, Abourehab MAS, El-Feky OA, et al. Biological effect of quercetin in repairing brain damage and cerebral changes in rats: Molecular docking and in vivo studies. BioMed Research International. 2022;**2022**:8962149. DOI: 10.1155/2022/8962149

[71] Harwood M, Danielewska-Nikiel B, Borzelleca JF, Flamm GW, Williams GM, Lines TC. A critical review of the data related to the safety of quercetin and lack of evidence of *in vivo* toxicity, including lack of genotoxic/carcinogenic properties. Food and Chemical Toxicology. 2007;**45**(11):2179-2205. DOI: 10.1016/j.fct.2007.05.015

[72] Dunnick JK, Halley JR. Toxicity and carcinogenicity studies of quercetin, a natural component of foods. Fundamental and Applied Toxicology. 1992;**19**(3):423-431. DOI: 10.1016/0272-0590(92)90181-g

[73] Okamoto T. Safety of quercetin for clinical application. International Journal of Molecular Medicine. 2005;**16**(2):275-278

[74] Pingili RB, Challa SR, Pawar AK, Toleti V, Kodali T, Koppula S. A systematic review on hepatoprotective activity of quercetin against various drugs and toxic agents: Evidence from preclinical studies. Phytotherapy Research. 2020;**34**(1):5-32. DOI: 10.1002/ptr.6503

[75] Chen R, Lin J, Hong J, Han D, Zhang AD, Lan R, et al. Potential toxicity of quercetin: The repression of mitochondrial copy number via decreased POLG expression and excessive TFAM expression in irradiated murine bone marrow. Toxicology Reports. 2014;**1**:450-458. DOI: 10.1016/j.toxrep.2014.07.014

[76] Dibal NI, Garba SH, Jacks TW. Acute toxicity of quercetin from onion skin in mice. Pharmaceutical and

Biomedical Research. 2020;**6**(4):269-276. DOI: 10.18502/pbr.v6i4.5113

[77] Sanchez-Gonzalez PD, Lopez-Hernandez FJ, Duenas M, Prieto M, Sanchez-Lopez E, Thomale J, et al. Differential effect of quercetin on cisplatin-induced toxicity in kidney and tumor tissues. Food and Chemical Toxicology. 2017;**107**:226-236. DOI: 10.1016/j.fct.2017.06.047

[78] Tiwari R, Siddiqui MH, Mahmood T, Farooqui A, Bagga P, Ahsan F, et al. An exploratory analysis on the toxicity & safety profile of Polyherbal combination of curcumin, quercetin and rutin. Clinical Phytoscience. 2020;**6**:82. DOI: 10.1186/s40816-020-00228-2

[79] Cunningham P, Patton E, VanderVeen BN, Unger C, Aladhami A, Enos RT, et al. Sub-chronic oral toxicity screening of quercetin in mice. BMC Complementary Medicine and Therapies. 2020;**22**(1):279. DOI: 10.1186/s12906-022-03758-z

[80] Mahdavinia M, Alizadeh S, Vanani AR, Dehghani MA, Shirani M, Alipour M, et al. Effects of quercetin on bisphenol A-induced mitochondrial toxicity in rat liver. Iranian Journal of Basic Medical Sciences. 2019;**22**(5):499. DOI: 10.22038/ijbms.2019.32486.7952

[81] Baran M, Yay A, Onder GO, Canturk Tan F, Yalcin B, Balcioglu E, et al. Hepatotoxicity and renal toxicity induced by radiation and the protective effect of quercetin in male albino rats. International Journal of Radiation Biology. 2022;**98**(9):1473-1483. DOI: 10.1080/09553002.2022.2033339

[82] Ruiz MJ, Fernández M, Estela JM, Asensi MÁ, Mañes J, Picó Y. Short-term oral toxicity of quercetin and pterostibene in Swiss mice. Toxicology Letters. 2006;**164**:S275-S276. DOI: 10.1016/j.toxlet.2006.07.232

[83] Utesch D, Feige K, Dasenbrock J, Broschard TH, Harwood M, Danielewska-Nikiel B, et al. Evaluation of the potential *in vivo* genotoxicity of quercetin. Mutation Research/Genetic Toxicology and Environmental Mutagenesis. 2008;**654**(1):38-44. DOI: 10.1016/j.mrgentox.2008.04.008

[84] Andres S, Pevny S, Ziegenhagen R, Bakhiya N, Schäfer B, Hirsch-Ernst KI, et al. Safety aspects of the use of quercetin as a dietary supplement. Molecular Nutrition & Food Research. 2018;**62**(1):1700447. DOI: 10.1002/mnfr.201700447

[85] Russo M, Spagnuolo C, Tedesco I, Bilotto S, Russo GL. The flavonoid quercetin in disease prevention and therapy: Facts and fancies. Biochemical Pharmacology. 2012;**83**(1):6-15. DOI: 10.1016/j.bcp.2011.08.010

[86] Kelly GS. Quercetin. Alternative Medicine Review. 2011;**16**(2):172-195

[87] Batiha GES, Beshbishy AM, Ikram M, Mulla ZS, El-Hack MEA, Taha AE, et al. The pharmacological activity, biochemical properties, and pharmacokinetics of the major natural polyphenolic flavonoid: Quercetin. Food. 2020;**9**(3):374. DOI: 10.3390/foods9030374

[88] Lakhanpal P, Rai DK. Quercetin: a versatile flavonoid. Internet Journal of Medical Update. 2007;**2**(2):22-37. DOI: 10.4314/ijmu.v2i2.39851

[89] Guillermo Gormaz J, Quintremil S, Rodrigo R. Cardiovascular disease: A target for the pharmacological effects of quercetin. Current Topics in Medicinal Chemistry. 2015;**15**(17):1735-1742. DOI: 10.2174/1568026615666150427124357

[90] Di Pierro F, Iqtadar S, Khan A, Ullah Mumtaz S, Masud Chaudhry M, Bertuccioli A, et al. Potential clinical benefits of quercetin in the early stage of COVID-19: Results of a second, pilot, randomized, controlled and open-label clinical trial. International Journal of General Medicine. 2021;**14**:2807-2816. DOI: 10.2147/IJGM.S318949

[91] Shen P, Lin W, Han L, Qin K, Huang Y, Tu S. Potential implications of quercetin in autoimmune diseases. Frontiers in Immunology. 2021;**12**:689044. DOI: 10.3389/fimmu.2021.689044

[92] Bhimanwar R, Kothapalli L, Khawshi A. Quercetin as natural bioavailability modulator: An overview. Research Journal of Pharmacy and Technology. 2020;**13**(4):2045-2052. DOI: 10.5958/0974-360X.2020.00368.6

[93] Erlund I, Kosonen T, Alfthan G, Mäenpää J, Perttunen K, Kenraali J, et al. Pharmacokinetics of quercetin from quercetin aglycone and rutin in healthy volunteers. European Journal of Clinical Pharmacology. 2000;**56**:545-553. DOI: 10.1007/s002280000197

[94] Riva A, Ronchi M, Petrangolini G, Bosisio S, Allegrini P. Improved oral absorption of quercetin from quercetin phytosome®, a new delivery system based on food grade lecithin. European Journal of Drug Metabolism and Pharmacokinetics. 2019;**44**:169-177. DOI: 10.1007/s13318-018-0517-3

[95] Rastogi H, Jana S. Evaluation of inhibitory effects of caffeic acid and quercetin on human liver cytochrome p450 activities. Phytotherapy Research. 2014;**28**(12):1873-1878

[96] Elbarbry F, Ung A, Abdelkawy K. Studying the inhibitory effect of quercetin and thymoquinone on human cytochrome P450 enzyme activities. Pharmacognosy Magazine. 2017;**13**(4):S895-S899. DOI: 10.4103/0973-1296.224342

[97] Östlund J, Zlabek V, Zamaratskaia G. *In vitro* inhibition of human CYP2E1 and CYP3A by quercetin and myricetin in hepatic microsomes is not gender dependent. Toxicology. 2017;**381**:10-18. DOI: 10.1016/j.tox.2017.02.012

[98] Wang Y, Cao J, Zeng S. Involvement of P-glycoprotein in regulating cellular levels of *ginkgo* flavonols: Quercetin, kaempferol, and isorhamnetin. Journal of Pharmacy and Pharmacology. 2005;**57**(6):751-758. DOI: 10.1211/0022357056299

[99] Wang SY, Duan KM, Li Y, Mei Y, Sheng H, Liu H, et al. Effect of quercetin on P-glycoprotein transport ability in Chinese healthy subjects. European Journal of Clinical Nutrition. 2013;**67**(4):390-394. DOI: 10.1038/ejcn.2013.5

[100] Borska S, Sopel M, Chmielewska M, Zabel M, Dziegiel P. Quercetin as a potential modulator of P-glycoprotein expression and function in cells of human pancreatic carcinoma line resistant to daunorubicin. Molecules. 2010;**15**(2):857-870. DOI: 10.3390/molecules15020857

[101] Kim MK, Choo H, Chong Y. Water-soluble and cleavable quercetin–amino acid conjugates as safe modulators for P-glycoprotein-based multidrug resistance. Journal of Medicinal Chemistry. 2014;**57**(17):7216-7233. DOI: 10.1021/jm500290c

[102] Ravikumar Reddy D, Khurana A, Bale S, Ravirala R, Samba Siva Reddy V, Mohankumar M, et al. Natural flavonoids silymarin and quercetin improve the brain distribution of

co-administered P-gp substrate drugs. Springerplus. 2016;**5**:1-9. DOI: 10.1186/s40064-016-3267-1

[103] Wang Y, Tao B, Wan Y, Sun Y, Wang L, Sun J, et al. Drug delivery based pharmacological enhancement and current insights of quercetin with therapeutic potential against oral diseases. Biomedicine & Pharmacotherapy. 2020;**128**:110372. DOI: 10.1016/j.biopha.2020.110372

[104] Caro C, Pourmadadi M, Eshaghi MM, Rahmani E, Shojaei S, Paiva-Santos AC, et al. Nanomaterials loaded with quercetin as an advanced tool for cancer treatment. Journal of Drug Delivery Science and Technology. 2022;**78**:103938. DOI: 10.1016/j.jddst.2022.103938

[105] Aghababaei F, Hadidi M. Recent advances in potential health benefits of quercetin. Pharmaceuticals. 2023;**16**(7):1020. DOI: 10.3390/ph16071020

[106] Liu Y, Huang F, Peterson BW, Zhou Q. Cutting-edge nanotechnologies in bacterial infection control. Frontiers in Chemistry. 2022;**10**:942370. DOI: 10.3389/fchem.2022.942370

[107] Bhoi A, Dwivedi SD, Singh D, Keshavkant S, Singh MR. Mechanistic prospective and pharmacological attributes of quercetin in attenuation of different types of arthritis. 3 Biotech. 2023;**13**(11):362. DOI: 10.1007/s13205-023-03787-6

[108] Shah N, Kumari M, Sadhu P, Talele C, Rajput HS, Sapra R. Nano-herbal innovations: Precision In therapeutic delivery. Journal of Advanced Zoology. 2024;**45**(1):576, 581. DOI: 10.17762/jaz.v45i1.2976

[109] Abo-zeid Y, Diab R, Sanad R, Sakran W. Recent advances in herbal-based nanomedicine for anti-inflammatory purposes. Journal of Advanced Pharmacy Research. 2021;**5**(4):387-397. DOI: 10.21608/APRH.2021.79946.1135

[110] Mahor A, Singh PP, Bharadwaj P, Sharma N, Yadav S, Rosenholm JM, et al. Carbon-based nanomaterials for delivery of biologicals and therapeutics: A cutting-edge technology. C-Journal of Carbon Research. 2021;**7**(1):19. DOI: 10.3390/c7010019

[111] Pinheiro RG, Pinheiro M, Neves AR. Nanotechnology innovations to enhance the therapeutic efficacy of quercetin. Nanomaterials. 2021;**11**(10):2658. DOI: 10.3390/nano11102658

[112] Wadhwa K, Kadian V, Puri V, Bhardwaj BY, Sharma A, Pahwa R, et al. New insights into quercetin nanoformulations for topical delivery. Phytomedicine Plus. 2022;**2**(2):100257. DOI: 10.1016/j.phyplu.2022.100257

[113] Tomou EM, Papakyriakopoulou P, Saitani EM, Valsami G, Pippa N, Skaltsa H. Recent advances in Nanoformulations for quercetin delivery. Pharmaceutics. 2023;**15**(6):1656. DOI: 10.3390/pharmaceutics15061656

[114] Zeng N, Chen X, Liu Z. Natural products and nanotechnology against coronavirus disease 2019. Frontiers in Chemistry. 2022;**10**:819969. DOI: 10.3389/fchem.2022.819969

[115] Mukherjee PK, Harwansh RK. Chapter 10 - Bioavailability of herbal products: Approach toward improved pharmacokinetics. In: Evidence-based Validation of Herbal Medicine. Amsterdam, Netherlands: Elsevier; 2015. pp. 217-245. DOI: 10.1016/B978-0-12-800874-4.00010-6

[116] Pearce K, Thipe VC, Henkel RR, Katti KV. Green nanotechnology as an innovative drug delivery approach for Typha capensis and Naringenin-new class of phytochemical embedded biocompatible gold nanoparticles in prostate cancer therapy. Journal of Drug Delivery Science and Technology. 2023;**80**:104100. DOI: 10.1016/j.jddst.2022.104100

[117] Díaz-Puertas R, Álvarez-Martínez FJ, Falco A, Barrajón-Catalán E, Mallavia R. Phytochemical-based nanomaterials against antibiotic-resistant bacteria: An updated review. Polymers. 2023;**15**(6):1392. DOI: 10.3390/polym15061392

[118] Teng H, Zheng Y, Cao H, Huang Q, Xiao J, Chen L. Enhancement of bioavailability and bioactivity of diet-derived flavonoids by application of nanotechnology: A review. Critical Reviews in Food Science and Nutrition. 2023;**63**(3):378-393. DOI: 10.1080/10408398.2021.1947772

[119] Khursheed R, Singh SK, Wadhwa S, Gulati M, Awasthi A. Enhancing the potential preclinical and clinical benefits of quercetin through novel drug delivery systems. Drug Discovery Today. 2020;**25**(1):209-222. DOI: 10.1016/j.drudis.2019.11.001

[120] Yang D, Liu X, Liu M, Chi H, Liu J, Han H. Protective effects of quercetin and taraxasterol against H2O2-induced human umbilical vein endothelial cell injury *in vitro*. Experimental and Therapeutic Medicine. 2015;**10**(4):1253-1260. DOI: 10.3892/etm.2015.2713

[121] Pippa N, Pispas S, Demetzos C. Polymer self-assembled nanostructures as innovative drug nanocarrier platforms. Current Pharmaceutical Design. 2016;**22**(19):2788-2795. DOI: 10.2174/1381612822666160217141232

[122] Chroni A, Mavromoustakos T, Pispas S. Biocompatible PEO-b-PCL nanosized micelles as drug carriers: Structure and drug-polymer interactions. Nanomaterials. 2020;**10**(9):1872. DOI: 10.3390/nano10091872

[123] Patel HS, Shaikh SJ, Ray D, Aswal VK, Vaidya F, Pathak C, et al. Formulation, solubilization, and *in vitro* characterization of quercetin-incorporated mixed micelles of PEO-PPO-PEO block copolymers. Applied Biochemistry and Biotechnology. 2022;**194**(1):445-463. DOI: 10.1007/s12010-021-03691-w

[124] Qi X, Gao C, Yin C, Fan J, Wu X, Di G, et al. Development of quercetin-loaded PVCL–PVA–PEG micelles and application in inhibiting tumor angiogenesis through the PI3K/Akt/VEGF pathway. Toxicology and Applied Pharmacology. 2022;**437**:115889. DOI: 10.1016/j.taap.2022.115889

[125] Elsabahy M, Wooley KL. Design of polymeric nanoparticles for biomedical delivery applications. Chemical Society Reviews. 2012;**41**(7):2545-2561. DOI: 10.1039/c2cs15327k

[126] Zhang Y, Chen J, Shi L, Ma F. Polymeric nanoparticle-based nanovaccines for cancer immunotherapy. Materials Horizons. 2023;**10**:361-392. DOI: 10.1039/D2MH01358D

[127] Sun L, Wu Q, Peng F, Liu L, Gong C. Strategies of polymeric nanoparticles for enhanced internalization in cancer therapy. Colloids and Surfaces B: Biointerfaces. 2015;**135**:56-72. DOI: 10.1016/j.colsurfb.2015.07.013

[128] Huang KT, Wu CT, Chang Y, Ho FM, Chiang CK, Liu SH. Therapeutic effect of quercetin polymeric nanoparticles on ischemia/reperfusion-induced acute

kidney injury in mice. Biochemical and Biophysical Research Communications. 2022;**608**:122-127. DOI: 10.1016/j.bbrc.2022.03.159

[129] Chaiwarit T, Chanabodeechalermrung B, Kantrong N, Chittasupho C, Jantrawut P. Fabrication and evaluation of water hyacinth cellulose-composited hydrogel containing quercetin for topical antibacterial applications. Gels. 2022;**8**(12):767. DOI: 10.3390/gels8120767

[130] Doosti M, Seyed Dorraji MS, Mousavi SN, Rasoulifard MH, Hosseini SH. Enhancing quercetin bioavailability by super paramagnetic starch-based hydrogel grafted with fumaric acid: An *in vitro* and *in vivo* study. Colloids and Surfaces B: Biointerfaces. 2019;**183**:110487. DOI: 10.1016/j.colsurfb.2019.110487

[131] Mok SW, Fu SC, Cheuk YC, Chu IM, Chan KM, Qin L, et al. Intra-articular delivery of quercetin using Thermosensitive hydrogel attenuate cartilage degradation in an osteoarthritis rat model. Cartilage. 2020;**11**(4):490-499. DOI: 10.1177/1947603518796550

[132] Gomes C, Dias R, Costa MR. Design of RAFT synthesized amphiphilic and stimuli-responsive block copolymers for encapsulation of polyphenols in polymersomes. In: 13th International Chemical and Biological Engineering Conference, Bragança, Portugal; 2018

[133] Shah S, Madhu H, Soniwala MM, Mori D, Vyas AJ, Chauhan A, et al. Lipid-based nanoparticles. In: Nanocarrier Vaccines: Biopharmaceutics-Based Fast Track Development. Hoboken, United States: Wiley-Scrivener; 2024. pp. 241-273. DOI: 10.1002/9781394175482.ch7

[134] Papadopoulou P, van der Pol R, van Hilten N, van Os WL, Pattipeiluhu R, Arias-Alpizar G, et al. Phase-separated lipid-based nanoparticles: Selective behavior at the nano-bio interface. Advanced Materials. 2024;**36**(6):e2310872. DOI: 10.1002/adma.202310872

[135] Patel P, Garala K, Singh S, Prajapati BG, Chittasupho C. Lipid-based nanoparticles in delivering bioactive compounds for improving therapeutic efficacy. Pharmaceuticals. 2024;**17**(3):329. DOI: 10.3390/ph17030329

[136] Akbarzadeh A, Rezaei-Sadabady R, Davaran S, Joo SW, Zarghami N, Hanifehpour Y, et al. Liposome: Classification, preparation, and applications. Nanoscale Research Letters. 2013;**8**(1):102

[137] Immordino ML, Dosio F, Cattel L. Stealth liposomes: Review of the basic science, rationale, and clinical applications, existing and potential. International Journal of Nanomedicine. 2006;**1**(3):297-315

[138] Munot NM, Shinde YD, Shah P, Patil A, Patil SB, Bhinge SD. Formulation and evaluation of chitosan-PLGA biocomposite scaffolds incorporated with quercetin liposomes made by QbD approach for improved healing of oral lesions. AAPS PharmSciTech. 2023;**24**(6):147. DOI: 10.1208/s12249-023-02584-x

[139] Dorostkar H, Haghiralsadat BF, Hemati M, Safari F, Hassanpour A, Naghib SM, et al. Reduction of doxorubicin-induced cardiotoxicity by co-administration of smart Liposomal Doxorubicin and free quercetin: *In vitro* and *in vivo* studies. Pharmaceutics. 2023;**15**(7):1920. DOI: 10.3390/pharmaceutics15071920

[140] Joshi H, Gupta DS, Kaur G, Singh T, Ramniwas S, Sak K, et al.

Nanoformulations of quercetin for controlled delivery: A review of preclinical anticancer studies. Naunyn-Schmiedeberg's Archives of Pharmacology. 2023;**396**(12):3443-3458. DOI: 10.1007/s00210-023-02625-z

[141] Liu D, Hu H, Lin Z, Chen D, Zhu Y, Hou S, et al. Quercetin deformable liposome: Preparation and efficacy against ultraviolet B induced skin damages in vitro and in vivo. Journal of Photochemistry and Photobiology B. 2013;**127**:8-17. DOI: 10.1016/j.jphotobiol.2013.07.014

[142] Shaji J, Iyer S. Preparation, optimization and *in-vivo* hepatoprotective evaluation of quercetin liposomes. International Journal of Current Pharmaceutical Research. 2012;**4**(2):24-32

[143] Yuan ZP, Chen LJ, Fan LY, Tang MH, Yang GL, Yang HS, et al. Liposomal quercetin efficiently suppresses growth of solid tumors in murine models. Clinical Cancer Research. 2006;**12**(10):3193-3199. DOI: 10.1158/1078-0432.CCR-05-2365

[144] Wong MY, Chiu GN. Liposome formulation of co-encapsulated vincristine and quercetin enhanced antitumor activity in a trastuzumab-insensitive breast tumor xenograft model. Nanomedicine: Nanotechnology, Biology and Medicine. 2011;7(6):834-840. DOI: 10.1016/j.nano.2011.02.001

[145] Priprem A, Watanatorn J, Sutthiparinyanont S, Phachonpai W, Muchimapura S. Anxiety and cognitive effects of quercetin liposomes in rats. Nanomedicine: Nanotechnology, Biology, and Medicine. 2008;**4**(1):70-78. DOI: 10.1016/j.nano.2007.12.001

[146] Phachonpai W, Wattanathorn J, Muchimapura S, Tong-Un T, Preechagoon D. Neuroprotective effect of quercetin encapsulated liposomes: A novel therapeutic strategy against Alzheimer's disease. American Journal of Applied Sciences. 2010;7(4):480-485. DOI: 10.3844/ajassp.2010.480.485

[147] Xu L, Wang X, Liu Y, Yang G, Falconer RJ, Zhao C. Lipid nanoparticles for drug delivery. Advanced NanoBiomed Research. 2022;**2**(2):2100109. DOI: 10.1002/anbr.202100109

[148] Paliwal R, Paliwal SR, Kenwat R, Kurmi BD, Sahu MK. Solid lipid nanoparticles: A review on recent perspectives and patents. Expert Opinion on Therapeutic Patents. 2020;**30**(3):179-194. DOI: 10.1080/13543776.2020.1720649

[149] Mirchandani Y, Patravale VB, Brijesh S. Solid lipid nanoparticles for hydrophilic drugs. Journal of Controlled Release. 2021;**335**:457-464. DOI: 10.1016/j.jconrel.2021.05.032

[150] Nguyen VH, Thuy VN, Van TV, Dao AH, Lee B. Nanostructured lipid carriers and their potential applications for versatile drug delivery via oral administration. OpenNano. 2022;**8**:100064. DOI: 10.1016/j.onano.2022.100064

[151] Li H, Zhao X, Ma Y, Zhai G, Li L, Lou H. Enhancement of gastrointestinal absorption of quercetin by solid lipid nanoparticles. Journal of Controlled Release. 2009;**133**(3):238-244. DOI: 10.1016/j.jconrel.2008.10.002

[152] Bose S, Du Y, Takhistov P, Michniak-Kohn B. Formulation optimization and topical delivery of quercetin from solid lipid based nanosystems. International Journal of Pharmaceutics. 2013;**441**(1-2):56-66. DOI: 10.1016/j.ijpharm.2012.12.013

[153] Sun M, Nie S, Pan X, Zhang R, Fan Z, Wang S. Quercetin-nanostructured lipid carriers: Characteristics and anti-breast cancer activities *in vitro*. Colloids and Surfaces B: Biointerfaces. 2014;**113**:15-24. DOI: 10.1016/j.colsurfb.2013.08.032

[154] Dhawan S, Kapil R, Singh B. Formulation development and systematic optimization of solid lipid nanoparticles of quercetin for improved brain delivery. Journal of Pharmacy and Pharmacology. 2011;**63**(3):342-351. DOI: 10.1111/j.2042-7158.2010.01225.x

[155] Moghtaderi M, Sedaghatnia K, Bourbour M, Fatemizadeh M, Salehi Moghaddam Z, Hejabi F, et al. Niosomes: A novel targeted drug delivery system for cancer. Medical Oncology. 2022;**39**(12):240. DOI: 10.1007/s12032-022-01836-3

[156] Yasamineh S, Yasamineh P, Ghafouri Kalajahi H, Gholizadeh O, Yekanipour Z, Afkhami H, et al. A state-of-the-art review on the recent advances of niosomes as a targeted drug delivery system. International Journal of Pharmaceutics. 2022;**624**:121878. DOI: 10.1016/j.ijpharm.2022.121878

[157] Carballo-Pedrares N, Kattar A, Concheiro A, Alvarez-Lorenzo C, Rey-Rico A. Niosomes-based gene delivery systems for effective transfection of human mesenchymal stem cells. Materials Science & Engineering. C, Materials for Biological Applications. 2021;**128**:112307. DOI: 10.1016/j.msec.2021.112307

[158] Shah S, Dhawan V, Holm R, Nagarsenker MS, Perrie Y. Liposomes: Advancements and innovation in the manufacturing process. Advanced Drug Delivery Reviews. 2020;**154-155**:102-122. DOI: 10.1016/j.addr.2020.07.002

[159] Elmowafy E, El-Derany MO, Biondo F, Tiboni M, Casettari L, Soliman ME. Quercetin loaded monolaurate sugar esters-based niosomes: Sustained release and mutual antioxidant-hepatoprotective interplay. Pharmaceutics. 2020;**12**(2):143. DOI: 10.3390/pharmaceutics12020143

[160] Javani R, Hashemi FS, Ghanbarzadeh B, Hamishehkar H. Quercetin-loaded niosomal nanoparticles prepared by the thin-layer hydration method: Formulation development, colloidal stability, and structural properties. LWT. 2021;**141**:110865. DOI: 10.1016/j.lwt.2021.110865

[161] Sadeghi-Ghadi Z, Ebrahimnejad P, Talebpour Amiri F, Nokhodchi A. Improved oral delivery of quercetin with hyaluronic acid containing niosomes as a promising formulation. Journal of Drug Targeting. 2021;**29**(2):225-234. DOI: 10.1080/1061186X.2020.1830408

[162] Sayyad N, Maji R, Omolo CA, Ganai AM, Ibrahim UH, Pathan TK, et al. Development of niosomes for encapsulating captopril-quercetin prodrug to combat hypertension. International Journal of Pharmaceutics. 2021;**609**:121191. DOI: 10.1016/j.ijpharm.2021.121191

[163] Wilson RJ, Li Y, Yang G, Zhao C. Nanoemulsions for drug delivery. Particuology. 2022;**64**:85-97. DOI: 10.1016/j.partic.2021.05.009

[164] Tayeb HH, Sainsbury F. Nanoemulsions in drug delivery: Formulation to medical application. Nanomedicine. 2018;**13**(19):2507-2525. DOI: 10.2217/nnm-2018-0088

[165] Arbain NH, Basri M, Salim N, Wui WT, Abdul Rahman MB. Development and characterization of aerosol nanoemulsion system

encapsulating low water soluble quercetin for lung cancer treatment. Materials Today: Proceedings. 2018;**5**:S137-S142. DOI: 10.1016/j.matpr.2018.08.055

[166] Mahadev M, Nandini HS, Ramu R, Gowda DV, Almarhoon ZM, Al-Ghorbani M, et al. Fabrication and evaluation of quercetin nanoemulsion: A delivery system with improved bioavailability and therapeutic efficacy in diabetes mellitus. Pharmaceuticals. 2022;**15**(1):70. DOI: 10.3390/ph15010070

[167] Nasirzadeh K, Nazarian S, Gheibi Hayat SM. Inorganic nanomaterials: A brief overview of the applications and developments in sensing and drug delivery. Journal of Applied Biotechnology Reports. 2016;**3**(2):395-402

[168] Mittal A, Roy I, Gandhi S. Magnetic nanoparticles: An overview for biomedical applications. Magnetochemistry. 2022;**8**(9):107. DOI: 10.3390/magnetochemistry8090107

[169] Elbeltagi S, Alfassam HE, Saeedi AM, Eldin ZE, Ibrahim EMM, Abdel shakor A, Abd El-Aal M. A novel quercetin-loaded NiFe2O4@ liposomes hybrid biocompatible as a potential chemotherapy/hyperthermia agent and cytotoxic effects on breast cancer cells. Journal of Drug Delivery Science and Technology. 2024;**91**:105203. DOI: 10.1016/j.jddst.2023.105203

[170] Barreto ACH, Santiago VR, Mazzetto SE, Denardin JC, Lavín R, Mele G, et al. Magnetic nanoparticles for a new drug delivery system to control quercetin releasing for cancer chemotherapy. Journal of Nanoparticle Research. 2011;**13**:6545-6553. DOI: 10.1007/s11051-011-0559-9

[171] Akal Z, Alpsoy L, Baykal A. Superparamagnetic iron oxide conjugated with folic acid and carboxylated quercetin for chemotherapy applications. Ceramics International. 2016;**42**(7):9065-9072. DOI: 10.1016/j.ceramint.2016.02.166

[172] Manzano M, Vallet-Regí M. Mesoporous silica nanoparticles for drug delivery. Advanced Functional Materials. 2020;**30**(2):1902634. DOI: 10.1002/adfm.201902634

[173] Frickenstein AN, Hagood JM, Britten CN, Abbott BS, McNally MW, Vopat CA, et al. Mesoporous silica nanoparticles: Properties and strategies for enhancing clinical effect. Pharmaceutics. 2021;**13**(4):570. DOI: 10.3390/pharmaceutics13040570

[174] Sapino S, Ugazio E, Gastaldi L, Miletto I, Berlier G, Zonari D, et al. Mesoporous silica as topical nanocarriers for quercetin: Characterization and *in vitro* studies. European Journal of Pharmaceutics and Biopharmaceutics. 2015;**89**:116-125. DOI: 10.1016/j.ejpb.2014.11.022

[175] Sarkar A, Ghosh S, Chowdhury S, Pandey B, Sil PC. Targeted delivery of quercetin loaded mesoporous silica nanoparticles to the breast cancer cells. Biochimica et Biophysica Acta. 2016;**1860**(10):2065-2075. DOI: 10.1016/j.bbagen.2016.07.001

[176] Poulson BG, Alsulami QA, Sharfalddin A, El Agammy EF, Mouffouk F, Emwas A-H, et al. Cyclodextrins: Structural, chemical, and physical properties, and applications. Polysaccharides. 2022;**3**(1):1-31. DOI: 10.3390/polysaccharides3010001

[177] Tian B, Liu J. The classification and application of cyclodextrin polymers: A review. New Journal of Chemistry. 2020;**44**(22):9137-9148. DOI: 10.1039/C9NJ05844C

[178] Jullian C, Moyano L, Yañez C, Olea-Azar C. Complexation of quercetin with three kinds of cyclodextrins: An antioxidant study. Spectrochimica Acta Part A: Molecular and Biomolecular Spectroscopy. 2007;**67**(1):230-234. DOI: 10.1016/j.saa.2006.07.006

[179] Liu M, Dong L, Chen A, Zheng Y, Sun D, Wang X, et al. Inclusion complexes of quercetin with three β-cyclodextrins derivatives at physiological pH: Spectroscopic study and antioxidant activity. Spectrochimica Acta. Part A, Molecular and Biomolecular Spectroscopy. 2013;**115**:854-860. DOI: 10.1016/j.saa.2013.07.008

[180] Kale R, Saraf M, Juvekar A, Tayade P. Decreased B16F10 melanoma growth and impaired tumour vascularization in BDF1 mice with quercetin-cyclodextrin binary system. The Journal of pharmacy and pharmacology. 2006;**58**(10):1351-1358. DOI: 10.1211/jpp.58.10.0008

[181] Kaşıkcı MB, Bağdatlıoğlu N. Bioavailability of quercetin. Current Research in Nutrition and Food Science Journal. (Special Issue Nutrition in Conference). 2016;**4**:146-151. DOI: 10.12944/CRNFSJ.4.Special-Issue-October.20

[182] Cai X, Fang Z, Dou J, Yu A, Zhai G. Bioavailability of quercetin: Problems and promises. Current Medicinal Chemistry. 2013;**20**(20):2572-2582. DOI: 10.2174/09298673113209990120

[183] Almeida AF, Borge GIA, Piskula M, Tudose A, Tudoreanu L, Valentová K, et al. Bioavailability of quercetin in humans with a focus on interindividual variation. Comprehensive Reviews in Food Science and Food Safety. 2018;**17**(3):714-731. DOI: 10.1111/1541-4337.12342

[184] Li Z, Deng H, Guo X, Yan S, Lu C, Zhao Z, et al. Effective dose/duration of natural flavonoid quercetin for treatment of diabetic nephropathy: A systematic review and meta-analysis of rodent data. Phytomedicine. 2022;**105**:154348. DOI: 10.1016/j.phymed.2022.154348

[185] Kagawad P, Gharge S, Jivaje K, Hiremath SI, Suryawanshi SS. Quality control and standardization of quercetin in herbal medicines by spectroscopic and chromatographic techniques. Future Journal of Pharmaceutical Sciences. 2021;7(1):176. DOI: 10.1186/s43094-021-00327-y

[186] Hollinger K, Shanely RA, Quindry JC, Selsby JT. Long-term quercetin dietary enrichment decreases muscle injury in mdx mice. Clinical Nutrition. 2015;**34**(3):515-522. DOI: 10.1016/j.clnu.2014.06.008

[187] Dian L, Yu E, Chen X, Wen X, Zhang Z, Qin L, et al. Enhancing oral bioavailability of quercetin using novel soluplus polymeric micelles. Nanoscale Research Letters. 2014;**9**(1):2406. DOI: 10.1186/1556-276X-9-684

[188] Kandemir K, Tomas M, McClements DJ, Capanoglu E. Recent advances on the improvement of quercetin bioavailability. Trends in Food Science & Technology. 2022;**119**:192-200. DOI: 10.1016/j.tifs.2021.11.032

[189] Manzoor MF, Hussain A, Sameen A, Sahar A, Khan S, Siddique R, et al. Novel extraction, rapid assessment and bioavailability improvement of quercetin: A review. Ultrasonics Sonochemistry. 2021;**78**:105686. DOI: 10.1016/j.ultsonch.2021.105686

[190] Miles SL, McFarland M, Niles RM. Molecular and physiological actions of quercetin: Need for clinical trials to assess its benefits in human disease.

Nutrition Reviews. 2014;**72**(11):720-734. DOI: 10.1111/nure.12152

[191] Han MK, Barreto TA, Martinez FJ, Comstock AT, Sajjan US. Randomised clinical trial to determine the safety of quercetin supplementation in patients with chronic obstructive pulmonary disease. BMJ Open Respiratory Research. 2020;**7**(1):e000392. DOI: 10.1136/bmjresp-2018-000392

[192] Csokay B, Prajda N, Weber G, Olah E. Molecular mechanisms in the antiproliferative action of quercetin. Life Sciences. 1997;**60**(24):2157-2163. DOI: 10.1016/s0024-3205(97)00230-0

[193] Kashyap D, Mittal S, Sak K, Singhal P, Tuli HS. Molecular mechanisms of action of quercetin in cancer: Recent advances. Tumor Biology. 2016;**37**:12927-12939. DOI: 10.1007/s13277-016-5184-x

[194] Suganthy N, Devi KP, Nabavi SF, Braidy N, Nabavi SM. Bioactive effects of quercetin in the central nervous system: Focusing on the mechanisms of actions. Biomedicine & Pharmacotherapy. 2016;**84**:892-908. DOI: 10.1016/j.biopha.2016.10.011

[195] Grewal AK, Singh TG, Sharma D, Sharma V, Singh M, Rahman MH, et al. Mechanistic insights and perspectives involved in neuroprotective action of quercetin. Biomedicine & Pharmacotherapy. 2021;**140**:111729. DOI: 10.1016/j.biopha.2021.111729

[196] Gupta A, Birhman K, Raheja I, Sharma SK, Kar HK. Quercetin: A wonder bioflavonoid with therapeutic potential in disease management. Asian Pacific Journal of Tropical Disease. 2016;**6**(3):248-252. DOI: 10.1016/S2222-1808(15)61024-6

[197] Oboh G, Ademosun AO, Ogunsuyi OB. Quercetin and its role in chronic diseases. Advances in Experimental Medicine and Biology. 2016;**929**:377-387. DOI: 10.1007/978-3-319-41342-6_17

[198] Shankar S, Kumar D, Srivastava RK. Epigenetic modifications by dietary phytochemicals: Implications for personalized nutrition. Pharmacology & Therapeutics. 2013;**138**(1):1-17. DOI: 10.1016/j.pharmthera.2012.11.002

[199] Schnarr L, Segatto ML, Olsson O, Zuin VG, Kümmerer K. Flavonoids as biopesticides–systematic assessment of sources, structures, activities and environmental fate. Science of the Total Environment. 2022;**824**:153781. DOI: 10.1016/j.scitotenv.2022.153781

[200] Mohi-ud-din R, Mir RH, Pottoo FH, Sawhney G, Masoodi MH, Bhat ZA. Nanophytomedicine ethical issues, regulatory aspects, and challenges. In: Nanophytomedicine. Singapore: Springer; 2020. pp. 173-192. DOI: 10.1007/978-981-15-4909-0_10

[201] Javaid M, Khan S, Haleem A, Rab S. Adoption of modern technologies for implementing industry 4.0: An integrated MCDM approach. Benchmarking: An International Journal. 2023;**30**(10):3753-3790. DOI: 10.1108/BIJ-01-2021-0017

[202] Cacciattolo M. Ethical considerations in research. In: Vicars M, Steinberg S, McKenna T, Cacciattolo M, editors. The Praxis of English Language Teaching and Learning (PELT), Critical New Literacies. Rotterdam, Netherlands: SensePublishers; 2015. DOI: 10.1007/978-94-6300-112-0_4

[203] Arifin SRM. Ethical considerations in qualitative study. International Journal of Care Scholars. 2018;**1**(2):30-33. DOI: 10.31436/ijcs.v1i2.82

[204] Abdel-Tawab M. Do we need plant food supplements? A critical examination of quality, safety, efficacy, and necessity for a new regulatory framework. Planta Medica. 2018;**84**(06/07):372-393. DOI: 10.1055/s-0043-123764

[205] Baenas N, Abellán Á, Rivera S, Moreno DA, García-Viguera C, Domínguez-Perles R. Foods and supplements. In: Polyphenols: Properties, Recovery, and Applications. Sawston, United Kingdom: Woodhead Publishing; 2018. pp. 327-362. DOI: 10.1016/B978-0-12-813572-3.00010-5

[206] Santini A, Cammarata SM, Capone G, Ianaro A, Tenore GC, Pani L, et al. Nutraceuticals: Opening the debate for a regulatory framework. British Journal of Clinical Pharmacology. 2018;**84**(4):659-672. DOI: 10.1111/bcp.13496

[207] Boots AW, Haenen GR, Bast A. Health effects of quercetin: From antioxidant to nutraceutical. European Journal of Pharmacology. 2008;**585**(2-3):325-337. DOI: 10.1016/j.ejphar.2008.03.008

[208] Vida RG, Fittler A, Somogyi-Végh A, Poór M. Dietary quercetin supplements: Assessment of online product informations and quantitation of quercetin in the products by high-performance liquid chromatography. Phytotherapy Research. 2019;**33**(7):1912-1920. DOI: 10.1002/ptr.6382

Section 2

Pharmacological Aspects

Chapter 3

Pharmacokinetics of Quercetin

Muhammad Candragupta Jihwaprani,
Wahyu Choirur Rizky and Mazhar Mushtaq

Abstract

Quercetin (QUE) is a primary polyphenol in the flavonoid family. It is categorized as one of the six subclasses of flavonoid compounds. As an abundant form of flavonoid molecules, quercetins are ubiquitously distributed in various dietary plants, including apples, berries, onions, bananas, tomatoes, and grapes. Furthermore, it is affordably marketed in the form of dietary supplement tablets. QUE is relatively lipophilic with low solubility in the water. Withal, QUE glucoside is more water soluble than the aglycone, and its absorption is limited to sodium-dependent glucose transporter-1 (SGLT-1); however, glucose transporter-2 (GLUT-2)-dependent absorption is also a significant contributor. Following absorption, QUE undergoes extensive metabolism in the liver, generating numerous metabolites. Data on the bioavailability of QUE differ substantially depending on the methods used for measuring QUE level. Pharmacokinetic interactions of QUE and its metabolites on cytochrome P450 enzymes have been studied extensively, but the results among the studies were inconsistent, such as weak inhibition toward CYP3A4 and no inhibition of CYP2D6 activity. Additionally, inhibition affects ATP- (adenosine triphosphate) binding cassette (ABC). Based on the pharmacokinetics profile, QUE has variable bioavailability based on the polymorphism of intestinal enzymes and transporters.

Keywords: quercetin, flavonoid, polyphenol, pharmacokinetic, antioxidant

1. Introduction

Quercetin (QUE) (3,3′,4′,5,7-pentahydroxyflavone) is a naturally occurring plant-based polyphenolic compound that belongs to the flavonols subclass of flavonoids. Within the flavonols subclass, other main compounds include isorhamnetin, myricetin, and kaempferol. Flavonoids, the primary secondary metabolites synthesized by various plants, play essential roles in plant defense and signaling in response to biological and environmental stressors [1]. Within plant cells, QUE and other flavonoids typically accumulate in the form of glycosides, the most common forms of which are QUE-3-glucoside and rutin [2]. Despite its abundance in various fruits and vegetables, its richest sources are onions, apples, grapes, berries, and tea [2, 3].

In recent years, there has been an increase in research interest in quercetin. QUE is the most studied phytochemical and an important flavonoid for research [4]. Specifically, a growing body of well-documented evidence suggests numerous therapeutic and prophylactic roles of QUE in various diseases (e.g., neurodegenerative, infectious, and cardiovascular) and other bioactive processes (e.g., antimicrobial, antiviral, antioxidants, and antiaging) [5, 6]. Most recently, numerous clinical trials have been conducted to examine the roles of QUE as an adjunct therapy for the coronavirus disease 2019 (COVID-19) [7].

Nowadays, QUE is also widely marketed in various pharmaceutical and dietary supplemental products in the form of capsules, tablets, and liquid drops. Some studies have examined the effect of incorporating QUE in different food products, for example, snack bars and chewing gums. More recently, QUE has been extensively studied for topical formulation, particularly in the nanocrystalline form, to enhance its dermal bioavailability [8, 9]. In line with this, it is essential to understand the pharmacokinetic properties of QUE in various formulations to determine the optimal dosage and carrier system that maximizes QUE uptake and its bioavailability [3, 10]. For instance, chewing bars, lozenges, and gums have certain advantages over traditional capsule formulations regarding rapid and efficient uptake in the bloodstream due to buccal absorption, thereby avoiding first-pass metabolism by the liver and intestinal cells. In this chapter, we will elaborate deeply on the pharmacokinetics properties of quercetin.

2. Dietary aspect of quercetin

QUE is a plant pigment classified as a flavonoid compound with a polyphenol or pentahydroxyflavone structure ($C_{15}H_{10}O_7$) [11]. Flavonoids are classified into six classes in relation to their chemical structure, including flavonols, flavanols, flavones, flavanones, isoflavones, and anthocyanidin [12]. It belongs to the class of flavonols that cannot be synthesized in the human body. Naturally, it is ubiquitous in plant food sources (primarily as glycosides) such as onions, apples, citrus fruits, green leafy vegetables, kale, broccoli grapes, cherries, berries, buckwheat, and green tea. Among them, onions and apples are the most essential sources of QUE in the human diet. Likewise, it has been considered a significant bioflavonoid compound in the human diet and has several benefits for human health [13]. Onions primarily contain QUE-4-glucoside and QUE-3,4′-diglucoside. On the other hand, apples contain QUE-3-O-glucoside, QUE-3-O-galactoside, QUE-3-O-rhamnoside, and QUE-3-O-rutinoside [14]. The amount of QUE contained in selected foods is presented in **Table 1**, and the chemical structure of QUE is illustrated in **Figure 1**.

Food Source	Quercetin Content (mg/100 g)
Capers	233.00
Onions	22.00
Cocoa powder	20.00
Cranberries	14.00
Lingonberries	7.40
Asparagus, cooked	7.61
Blueberries	5.05
Apple, red	4.70
Cherries	2.64
Broccoli, raw	2.51

Food Source	Quercetin Content (mg/100 g)
Apple, Fuji	2.02
Green tea	2.69
Black tea	1.99
Red grapes	1.38

Table 1.
Amount of quercetin in selected foods [15].

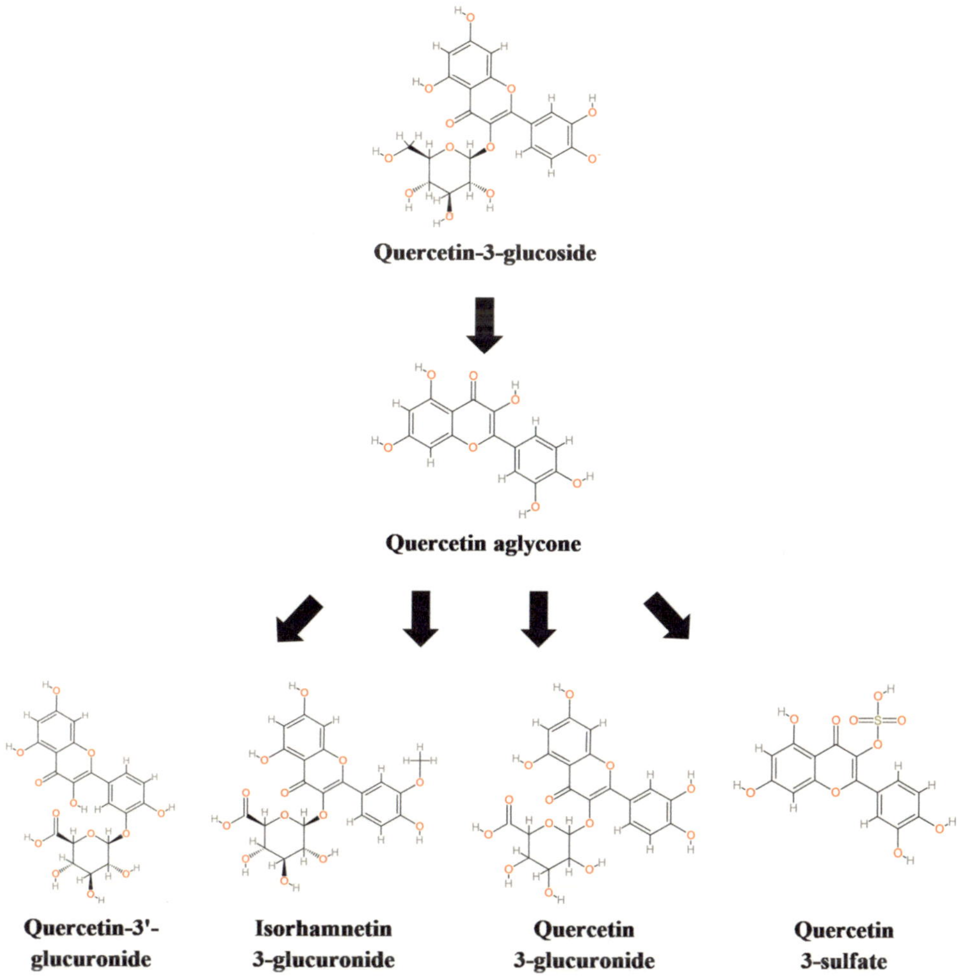

Figure 1.
Chemical structure of quercetin and its metabolites.

In contrast to naturally occurring QUE, mainly in the form of glycoside, dietary supplements in the marketplace containing QUE are provided as a free form of quercetin the aglycone [14]. Pharmacologically, QUE has been evaluated for its antiviral and antibacterial activity, anti-inflammatory effects, antiplatelets, antihypertensive, antitumor, neuroprotective effect, cardioprotective effect, gastroprotective effect, natural antihistamine, hepatoprotective, and antiprotozoal [12, 13, 16]. In Western diets, intake of QUE is between 3 and 40 mg per day as aglycones equivalent. In "high-end fruits and vegetables consumers," the intake has been estimated equal to 250 mg per day [14]. Intended daily doses of QUE as a dietary supplement are often considerably higher than QUE levels taken from natural dietary products (i.e., fruits and vegetables). Most commonly, the recommended daily dose of QUE as a form of dietary supplement (QUE aglycone) is equal to 500 mg (maximum dose is 1000 mg) [14]. Some manufacturers in some countries recommend daily doses of 200 to 1200 mg QUE [17].

Identification of different QUE in any food is possible with different types of chromatography; some of these bioassays are an excellent choice for the quantitative and qualitative analysis of various metabolites. By these methods, a group of scientists has successfully shown the extraction of QUE and other flavonoids in the mulberry leaf, apples, and onions, which are the potential sources of QUE [18, 19].

3. Absorption of quercetin

3.1 QUE is deglycosylated prior to absorption

Most of the QUE in food plants is bound to a sugar molecule *via* a beta-glycosidic link (also known as glycosyl group), and this conjugation is called glycoside. The factors that mainly affect QUE absorption include the nature of the attached sugar and the solubility as modified by ethanol, fat, and emulsifiers [20]. Sugar molecules are released from the glycosides after mastication, digestion, and absorption [13]. Prior to absorption, QUE glycosides (QUE glucoside, QUE arabinoside, and QUE galactoside) are deglycosylated to QUE aglycone by lactase phlorizin hydrolase (LPH) and cytosolic β-glycosidases (CBG) in the enterocytes brush border [21, 22]. The deglycosylation of QUE glycosides possesses an essential role in increasing its absorption in the intestine, increasing its plasma concentration, and improving its bioavailability [23]. It is established that QUE aglycones are more readily absorbed due to their relatively higher lipophilicity in contrast to their glycoside counterparts, where the absorption *in vivo* is significant through passive diffusion (**Figure 2**) [24, 25].

As LPH works directly on the intestinal lumen by cleaving polar glycosides, the released aglycones are then able to passively diffuse across the intestinal wall [22]. However, LPH is not evenly distributed and expressed along the gastrointestinal tracks (GIT) of mammals, mainly because of region specificity and the postweaning decline. Considering that the enzyme in the enterocytes brush border is specific for glucose, QUE glucosides are absorbed more quickly than other glycosides, such as rutin (QUE-3-O-rutinoside) [26]. Rutin (QUE-3-O-rutinose) is neither the substrate of LPH nor CBG, which is hardly absorbed in the small intestine and delivered to the distal part of the GIT, where it is absorbed following degradation by colonic microbiota [27]. Fascinatingly, in the lower gut, the deglycosylation of rutin (QUE-3-O-rutinoside) is mediated by CBG secreted by the gut microbiota or microbial hydrolases (by α-rhamnosidase and β-glucosidase) instead of that by the colonic epithelium since LPH and CBG expression in the latter is insignificant and low [28]. Hence, the absorption rate and efficiency of rutin are considerably less in contrast to other glycosides [29, 30]. In light of this, an *in vitro* study revealed that approximately 60% of QUE rutinoside were degraded to 3,4-dihydroxyphenylacetic acid within 2 hours by the microbiota in the colon, proposing that most QUE rutinoside is initially

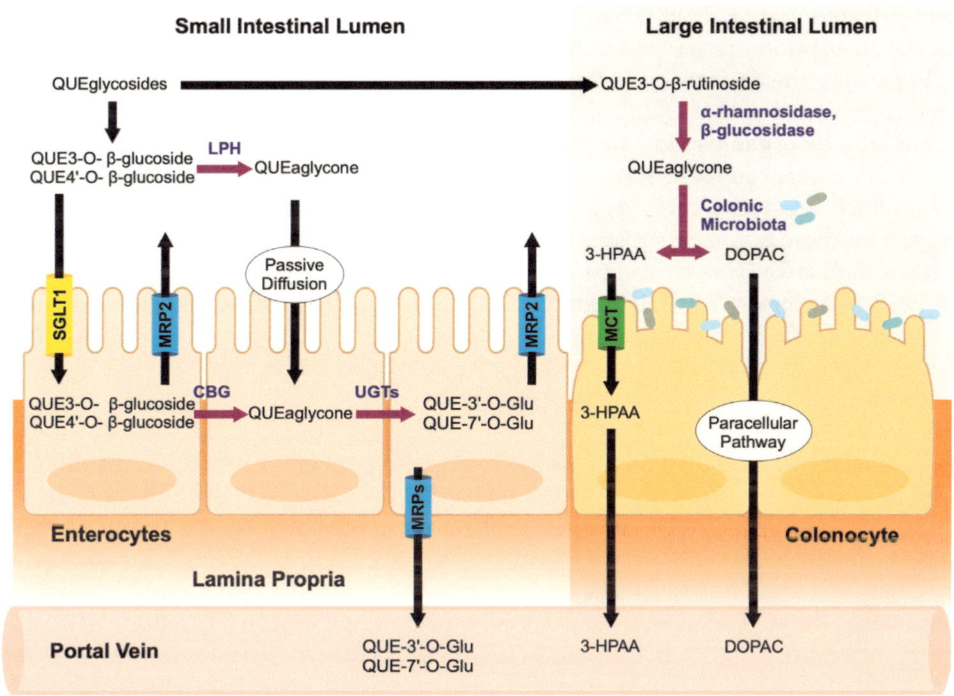

Figure 2.
Processing of quercetin glycosides in small intestine and large intestine. 3-HPAA, 3-hydroxyphenylacetic acid; CBG, β-cytosolic glucosidases; DOPAC, 3,4-dihydroxyphenylacetic acid; Glu, glucuronide; LPH, lactase-phlorizin hydrolase; MCT, monocarboxylate transporter; MRP2, multiresistance protein 2; QUE, quercetin; and UGT, uridine-5′-diphosphate glucuronosyl transferases.

deglycosylated to QUE aglycone before its degradation to 3,4-dihydroxyphenylacetic acid and 3-hydroxyphenylacetic acid [20].

Moreover, QUE-3-glucoside and rutin have been detected in the basolateral membrane monolayer *in vitro*, detected and identified in plasma in *in-vivo* studies. Some studies also reported that all forms of polyphenols, including intact aglycones, the original glycosides, and the metabolites, coexist in fecal samples in the colon [28]. Therefore, the absorption mechanism in the gastrointestinal tract, especially quercetin, must be revisited.

3.2 Specific membrane transporter mediates QUE absorption

QUE conjugates are challenging to pass *via* cellular membrane because a lipid bilayer with high membrane polarity requires a specific transporter to cross the membrane. QUE is relatively lipophilic with low solubility in the water. However, QUE glucoside is more water soluble than the aglycone [21]. Additionally, QUE glucoside is absorbed *via* sodium-dependent glucose transporter-1 (SGLT-1) but not with QUE aglycone. Hence, SGLT-1-mediated absorption contributes to a tremendous amount of the glucoside absorption [21, 31]. Glucose transporter 2 (GLUT-2) also plays a role [31]. The pathway and site of QUE absorption rely on its chemical structure. Preclinical studies hypothesize that the glucose moiety utilizes a transporter to aid its absorption across the intestinal lumen. Previous studies using rat models demonstrated that QUE aglycone is absorbed both in the stomach and the intestine, but the absorption mechanism in the stomach remains unclear.

Regarding the absorption in the intestine, some studies utilizing the caco-2 cell monolayer revealed that QUE aglycone is absorbed primarily by passive diffusion and secondarily by organic anion-transporting polypeptides (OATPs) [21, 32]. When flavonols are present in the dietary form of aglycone, they could be partially absorbed in the stomach relative to their glycoside forms, which are not absorbed. The absorption of QUE by these two mechanisms is pH dependent, in which in lower pH, QUE exerts a high affinity as a substrate of OATP-B uptaken into enterocytes *via* the OATP-B transporter. While in higher pH, QUE is absorbed through a passive diffusion mechanism [32].

3.3 Factors influencing quercetin absorption

The absorption of QUE, such as other flavonoids, is considered poor, resulting in its variably limited bioavailability [29]. QUE absorption is influenced by various factors, including its glycosylation, preparation, coadministration of dietary components along with QUE, and interindividual variability [26, 33]. Preclinical studies demonstrated that QUE glucoside had a greater bioavailability compared to aglycone, presumably due to differences in absorption properties [33]. In particular, QUE glucosides are biochemically more hydrophilic than aglycones. Specific transporters also exclusively facilitate QUE glucoside absorption, including SGLT-1, which contributes to higher intestinal uptake of glucosides (**Figure 2**) [33, 34]. The sugar moiety in the glucosides also enhances its solubility to aid in absorption. Similar studies conducted in healthy subjects with ileostomy (to avoid the role of losses by colonic bacteria) demonstrated higher absorption in glucosides from onions than pure aglycone (52 ± 15 vs. $24 \pm 9\%$) [35]. QUE bioavailability is also better when prepared in cereal bars than in capsule forms due to the homogenous solid dispersion of QUE with other components in the cereal, promoting dissolution in the intestinal lumen [36].

Co-ingestion of QUE with other dietary nutrients also affects its absorption. Dietary fat enhanced proved the absorption of QUE in a randomized crossover study among overweight/obese and menopausal subjects, improving its plasma maximum concentration (C_{max}) and the area under the plasma concentration-time curve (AUC) over the 24 hours by 45 and 32% compared to fat-free subjects, respectively. Dietary fats may enhance QUE absorption by micellization in small intestines [37]. Coadministration of QUE with fructooligosaccharides (found naturally in various plants) also increases its bioavailability in preclinical studies [38]. Piperine (Pip), a compound responsible for the pungency of black pepper, is also known to enhance QUE absorption and bioavailability. It improves GIT absorption *via* multiple mechanisms, including augmenting compound solubility, increasing epithelial cell permeability, and increasing intestinal blood supply. It is also a potent inhibitor of drug hepatic metabolism, thus enhancing QUE bioavailability [39]. A study examining the effect of adding pip in QUE-loaded nanosuspensions demonstrated significantly greater absolute oral bioavailability (23.58% pip-containing nanosuspensions vs. 3.61–15.55% in other formulations) [40]. Vitamin C may also augment QUE bioavailability [33]. From the absorption point of view, ascorbic acid, especially at a high dose, may increase intestinal permeability and stability of QUE molecules to become more soluble in the GIT [41]. However, the evidence demonstrating this effect in humans is lacking. Similarly, QUE coadministration with bromelain, a crude pineapple extract, also enhanced QUE's oral bioavailability by up to 80% [42].

On the other hand, some dietary companions may decrease QUE bioavailability. For instance, iron, especially nonheme iron in plant-based foods and various supplements, can bind to QUE and reduce absorption. Similarly, QUE and other flavonoids also significantly mitigate iron absorption due to the inhibition of basolateral transport across intestinal epithelial cells [43]. Calcium supplements may also interfere with QUE absorption despite scarce evidence. The mechanism underlying its inhibitory effects is theoretical because calcium can form insoluble complexes with QUE in the GIT, thus reducing its bioavailability. Certain medications may interfere with the absorption and bioavailability of QUE and will be further discussed in the subsequent section (see Section 6. Drug Interactions of Quercetin).

4. Bioavailability of quercetin

Bioavailability refers to a ratio between the amount of orally ingested substance and the amount that is absorbed and completely available to its intended biological destinations and to exert physiologic activity or storage [11]. For polyphenols, this is often deemed as the amount detected in plasma. The minimum bioavailability can also be calculated as a percentage based on the urinary measurement of the compound and its metabolites [26]. Generally, bioavailability is calculated by measuring QUE derivatives or catabolites in blood or urine [26]. Given that QUE is primarily detected in several conjugated forms *in vivo*, the analytical procedures followed by most authors are animal and human studies. It is shown that QUE had poor oral bioavailability after a single oral dose, presumably due to variability in the absorption rates related to polymorphism of intestinal enzymes and transporters [45]. The bioavailability of QUE is associated with its bioaccessibility and, thus, solubility in the vehicle for administration. Crystalline formation at body temperature and the poor solubility of QUE restrict its bioavailability and bioaccessibility [21]. Likewise, QUE glycosides or aglycone are effluxed back across the apical membrane into the intestinal lumen following its enterocyte uptake [46]. Data on the bioavailability of QUE (glycoside and aglycone) differ substantially between studies depending on the methods used for measuring QUE level. It also varies between each species. For instance, in human studies, interindividual variations could be influenced by the complexity of QUE absorption and metabolism, dietary adaptation, genetic polymorphism, body mass index (BMI), the composition of gut microflora, and drug-drug interactions [26]. The bioavailability of QUE can be improved by ingestion with short-chain fructooligosaccharide (FOS) or form of cereal bar ingredient rather than in capsule form [21]. Co-ingestion with dietary lipids increases the intestinal absorption of quercetin. A previous clinical study by Guo *et al.* suggested that individual differences in the level of plasma vitamin C may contribute to intersubject variability in the bioavailability of QUE [47]. Additionally, some *in vitro* studies revealed that the presence of vitamin C can protect QUE against oxidative degradation [34], yet further studies are required regarding the role of vitamin C in regulating the bioavailability of quercetin.

In one clinical study, 35 healthy subjects were randomly assigned to take a designated dose of QUE capsules of either 50, 100, or 150 mg daily for 2 weeks. The supplementation significantly raised plasma levels of QUE by 178, 359, and 570% for each dose of 50, 100, and 150 mg, respectively, compared to baseline status. The AUC was between 76.1 and 305.8 mmol/min/L (50- and 150-mg dosages, respectively). The achieved median maximum plasma level was approximately 431 nmol/L after

360 minutes of ingesting 150 mg of quercetin. Based on this study, daily supplementation of QUE for 2 weeks dose-dependently raised plasma levels of QUE without affecting antioxidant status, inflammation, metabolism, and oxidized LDL [17]. Another larger randomized placebo-controlled trial involving 1002 healthy subjects showed that QUE supplementation in doses of 500 and 1000 mg per day significantly raised plasma levels of QUE but was highly variable after continuous administration over 12 weeks. However, the variability was unrelated to the age, gender, BMI, and lifestyle factors of the subjects [45].

In clinical studies, the extent of QUE absorption can be estimated by multiplying the C_{max} by plasma volume (on average 3.5 L in a normal 70-kg adult) and dividing by the administered dose of quercetin. Likewise, scanty amounts of QUE aglycone were detected in plasma compared to QUE conjugates (glycoside), which could be recovered in plasma after oral ingestion [14]. The pharmacokinetics of intravenous QUE in cancer patients were studied by Ferry et al. at dose levels of 60–2000 mg/m^2 [48]. It was concluded in this study that 945 mg/m^2 was the safe intravenous dose of quercetin. In the same survey, a toxic amount was reported to induce emesis, hypertension, nephrotoxicity, and decreased serum potassium. Ferry et al. further concluded that the distribution and elimination half-life ($t^{1/2}$) of intravenous QUE was 0.7–7.8 and 3.8–86 min, respectively, and its clearance and distribution volume was 0.23–0.84 and 3.7 L/m^2, respectively [29, 30, 48].

5. Metabolism of quercetin

Recently, hepatic transport processes have been recognized as an essential determinant of drug disposition. Therefore, unsurprisingly, the characterization of potential drug candidates' hepatic transport and biliary excretion properties is necessary for the drug development process [49]. Basolateral transport systems are responsible for translocating molecules across the sinusoidal membrane. Conversely, active canalicular transport systems are responsible for the biliary excretion of drugs and metabolites [50]. Numerous transport proteins involved in basolateral and canalicular transport have been identified. QUE has been proven to be the substrate of OATP2B1, OATP1A2, and organic cation transporter-1 (OCT1) in the OATPs-HEK293 cell line. In addition to passive diffusion, OATP2B1, OATP1A2, and OCT1 may also arbitrate QUE transport and contribute to QUE accumulation in hepatocytes [51]. Moreover, OAT4 on the basolateral membrane of hepatocytes confers an essential role in the cellular uptake of QUE-3′-sulfate [52]. On the one hand, the primary intestinal metabolites of quercetin, QUE glucuronides, may be transported primarily by passive diffusion [53].

QUE metabolism occurs mainly in the gastrointestinal tract. In mouse models, approximately 93% of QUE aglycone, or the one from hydrolyzed derivatives, is metabolized in the intestine after absorption. Albeit most QUE is metabolized in the small intestine, the liver contains all the enzymatic systems that allow its complete metabolism through reactions of methylation, sulfation, and glucuronide conjugation. Efficient glucuronidation of QUE takes place in enterocytes by UDP-glucuronyltransferases (UGT), methylation by catechol-O-methyltransferases (COMT), and sulfation by sulfotransferases (SULT) [54]. In an attempt to migrate from the cell to the bloodstream, conjugation has to occur, and conjugated QUE metabolites are formed. Consequently, QUE aglycone is present in low amounts in plasma. The two principal metabolites in humans that pass into the blood from the enterocyte are QUE-3-O-glucuronide and QUE 3′-O-sulfate, which are transported from the enterocyte to the liver via the portal vein [55, 56].

QUE and its metabolites will undergo phases I and II metabolism while reaching the liver, similar to other flavonoids. Phase I metabolism mainly includes oxidation, reduction, and hydrolysis by facilitating its subsequent metabolism by introducing a physiological functional group. Hence, it can increase the reactivity of the substrate. Phase I metabolism is mediated by cytochrome (CYP) 450 (CYP450) and has minimal effect on overall flavonoid metabolism [57]. However, QUE can affect the substrate bioavailability of CYPs by activating or inhibiting CYP activity. QUE has been identified as a competitive inhibitor of CYP2C19, CYP3A4, and CYP2D6 in human liver microsomes model [58].

QUE and its metabolites undergo further phase II metabolism in the liver following specific pathways: glucuronidation, sulfation, and methylation. These pathways facilitate the excretion of QUE and its metabolites through bile and urine [59]. Glucuronide conjugates are the primary existing forms of QUE in the bloodstream. UGTs promote the conversion of hydrophobic substrates to hydrophilic glucuronides, which are vital in reducing the bioactivities of xenobiotics [60]. Sulfation has the primary role of reducing potential toxicity by adding sulfate groups to xenobiotics. The 3′-OH is reckoned to be an essential site for the sulfation of quercetin. COMT mediates the methylation of QUE at 3′-OH or 4′-OH of catechol to generate 3′O-methylquercetin and 4′OH-methylquercetin [48], respectively. Glucuronide derivatives are the primary metabolites of QUE in the intestine and liver. QUE-3′-sulfate was also found as another significant metabolite. The generation of QUE sulfate conjugate shows that QUE glucuronide conjugates can still be mediated by β-glucuronidase in the liver to generate QUE aglycone. SULT then metabolizes QUE aglycone to form the corresponding sulfate [61].

Following metabolism in the liver, the QUE metabolites are secreted into bile and appear in the feces. Approximately 35% of QUE metabolites were observed in bile, and metabolites resulting from glucuronidation and sulfation were the predominant existing forms [62]. As seen in **Figure 3**, QUE phase II metabolites are the substrates of breast cancer resistance protein (BCRP) and multiresistance protein 2 (MRP2) efflux transporter. Specifically, MRP2, located on the apical membrane of hepatocytes, facilitates the secretion of QUE glucuronide and QUE 3′-O-sulfate into bile, leading to a lower bioavailability of QUE metabolites [61]. Synchronously, QUE glucuronide and sulfate conjugates secreted into bile and reaching the intestine may then be reabsorbed and thus enter the portal vein circulation [63]. Nevertheless, no significant absorption of QUE and its conjugates was identified in mouse models by intravenous injection of QUE and its bile into the duodenum, denoting QUE metabolites that do not possess enterohepatic circulation [64]. Hence, we require further studies to understand whether enterohepatic circulation can improve the bioavailability of quercetin.

Additionally, phase II conjugates of QUE in the liver may also be transported to the circulation through MRPs located on the basolateral membrane of hepatocytes [65]. QUE-3-O-glucuronide, 3′-methylquercetin-3-O-glucuronide, and QUE 3′-O-sulfate are the main existing forms of QUE in human plasma after oral onions. In plasma, QUE metabolites bind to albumin primarily via non-covalent bonding, and the binding ratio is approximately 70–80% distinct in the antioxidant activity between various QUE conjugates [66]. Contrary to QUE aglycone, the biological activity of phase II conjugates is reduced, such as methylated QUE with reduced antioxidant activity [67].

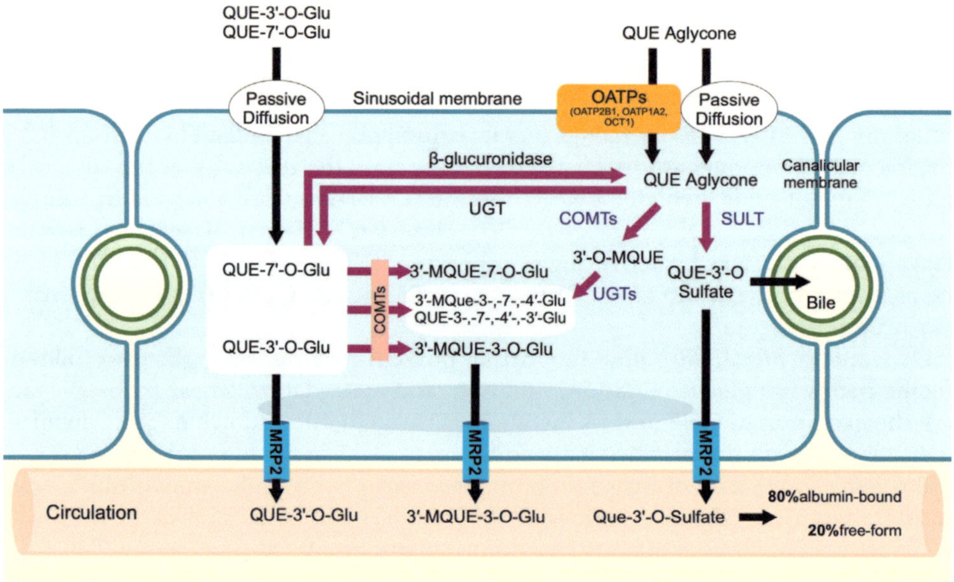

Figure 3.
The process of metabolism and transport of QUE and its metabolites in the liver. COMTs, catechol-O-methyltransferases; Glu, glucuronide; MQUE, methyl quercetin; MRP2, multi-resistance protein 2; OATPs, organic anion transport polypeptides; OCT, organic cation transporter; QUE, quercetin; SULT, sulfotransferases; UGTs, uridine-5′-diphosphate glucuronosyl transferases.

6. Drug interactions of quercetin

A crossover clinical study demonstrated that single-dose and repeated intake (-short-term) of QUE (1500 mg/day) decreased the C_{max} and AUC0-∞ values of talinolol (substrate of intestinal P-glycoprotein) in human volunteers. The authors suggested that this occurs due to interaction between quercetin, P-glycoprotein (efflux), and OATP (uptake) transporters, predominantly by inhibition of talinolol absorption mediated by OATP [68].

Pharmacokinetic interactions of QUE aglycone and its primary metabolites with the albumin binding of warfarin were also evaluated *in vitro*. A study by Poor et al. [69] showed that QUE metabolites exhibited high affinity binding to human serum albumin, and the methylated and sulfated metabolites had similar or even greater abilities to displace warfarin binding to albumin than the parent compound. On the other hand, the glucuronide conjugates exhibited lower affinity binding to human serum albumin; thus, the competition with warfarin was also lower. The researchers suggested that repeated exposure to large doses of QUE renders significant interaction with warfarin since the displacement of warfarin from human serum albumin leads to bleeding, and subsequent serious effects may follow. However, the study showed that QUE and its metabolites exerted a considerably lower impact on hydroxylation catalyzed by CYP2C9 than warfarin [69].

Contrariwise, a randomized, open-label crossover study by Nguyen et al. [70] demonstrated that a single oral dose of 1500 mg QUE did not alter midazolam pharmacokinetics significantly, while one-week supplementation of QUE rendered a trend to decreased midazolam exposure and reduced the ratio of midazolam - 1′-hydroxymidazolam $AUC_{0-\infty}$. It indicated that QUE induced a CYP3A4 activity.

Likewise, the coadministration of QUE and midazolam would not trigger any toxic adverse effects. However, the therapeutic efficacy of midazolam may be reduced when administered orally following a short-term high-dose QUE supplementation, presumably due to elevated CYP3A4-mediated metabolism of the drug [70]. Another crossover study demonstrated that concomitant short-term intake of QUE (500 mg three times a day for 1 week) significantly increased the mean plasma concentration of fexofenadine compared to those of the placebo phase. However, there was a significant decrease of 37% in oral clearance of fexofenadine after QUE intake. There was no difference in the t½ and renal clearance between the QUE and placebo phases. It was suggested that inhibition of p-glycoprotein-mediated efflux was involved in this process [71].

Individuals on therapeutic regimens with narrow therapeutic windows should be under physician supervision before use. It has been shown that QUE enhances antifungals and antibiotics, particularly those targeting the fluconazole-resistant group of *Candida albicans* [72] and *Candida tropicalis* [73], multidrug-resistant *Pseudomonas aeruginosa* [74], amoxicillin-resistant *Staphylococcus epidermidis* [75], and multidrug-resistant *Escherichia coli* [76]. It also competitively inhibits bacterial DNA gyrase; hence, it is contraindicated to concomitant use with fluoroquinolones (ciprofloxacin, levofloxacin, moxifloxacin, and ofloxacin) [77].

7. Excretion of quercetin

Apart from the transport and excretion of the metabolized QUE in its glucuronidated, methylated, and sulfated forms into the bile (**Figure 3**), a considerable amount of its metabolites are transported via the bloodstream into the kidneys. Furthermore, in addition to exerting its renoprotective effects (e.g., antioxidant, anti-inflammatory, anti-fibrotic), kidneys would also transport the metabolites into the renal cells and renal tubules for excretion. Some hydrophilic metabolites from the bloodstream may go through glomerular filtration, while others are secreted into the tubule, mainly in the proximal convoluted tubule (PCT) [54, 78].

The transport of QUE to PCT epithelial cells principally occurs in the basolateral membrane (BLM) and brush border membrane (BBM) [79]. OATs are abundantly expressed on the BLM of the PCT epithelial cells to facilitate the transport. Significantly, QUE 3′-O-sulfate and QUE 3′-O-glucuronide are actively transported into the PCT cells owing to high affinity for OAT1 and OAT3, respectively (**Figure 4**) [46]. Contrarily, QUE 3-O-glucuronide and QUE 7-O-glucuronide have poor affinity for OAT1 and OAT3. On the side of BBM of PCT epithelial cells, MRP2 and BCRP are the primary transporters responsible for the secretion of glucuronidated QUE, whereas its sulfated metabolites are transported *via* MRP2 [46, 61].

QUE and its metabolites are also partially reabsorbed into the tubular cells [78]. QUE aglycone and methyl-quercetin can be passively reabsorbed into the tubular cells across the BBM, whereas its sulfated and glucuronidated conjugates necessitate active transporters for reabsorption. The passive influx of aglycone across BBM is thought to be due to its lipophilic properties, whereas methylated conjugates can be passively reabsorbed owing to their planar configuration. On the other hand, most hydrophilic metabolites are less likely to be reabsorbed under physiological circumstances. However, some transporters for active reabsorption have been identified and may play a role in active transport across BBM, which include organic anion transporter 4 (OAT4) for sulfated conjugates and MRP2 and BRCP for glucuronides [80].

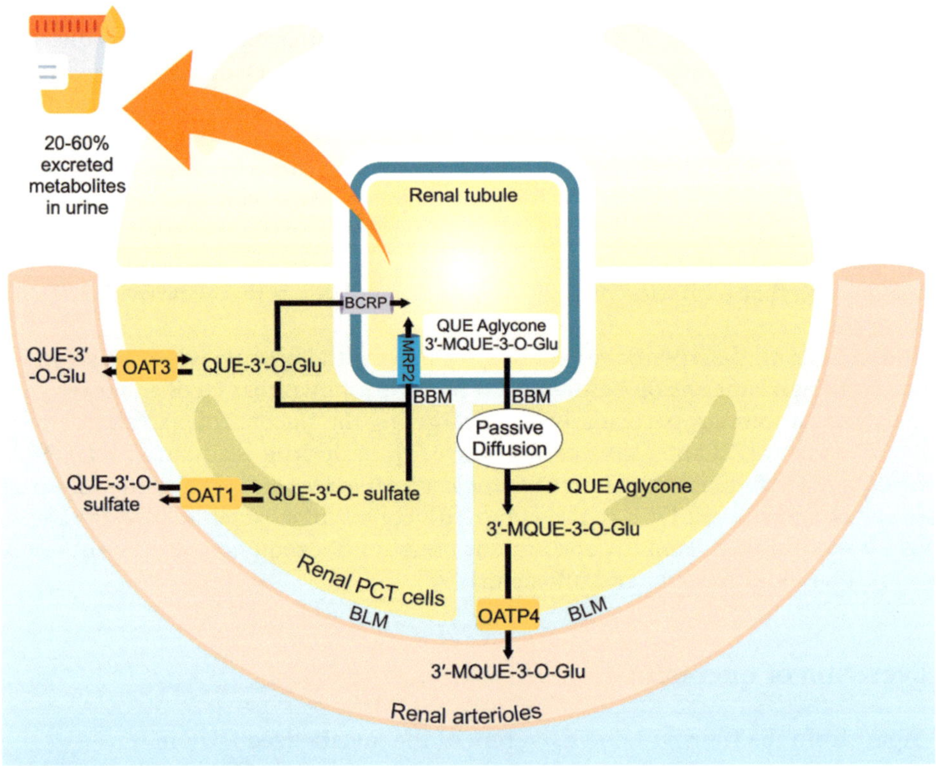

Figure 4.
Schematic illustration of the transport and excretion of quercetin by the kidney. BCRP, breast cancer receptor protein; BLM, basolateral membrane; BMM, brush border membrane; Glu, glucuronide; MQUE, methyl-quercetin; MRP2, multi-resistance protein 2; OAT, organic anion transporter; OATP, organic anion transport polypeptide; PCT, proximal convoluted tubule; and QUE, quercetin.

Mullen et al. examined the excreted metabolites of QUE after ingesting lightly fried onions in healthy human subjects. They concluded that the proportion of QUE metabolites being excreted in the urine from the dietary intake of QUE is around 4.7% of the total intake (12.9 μmol) [81]. In addition, the profile of metabolites excreted in urine differed significantly from that of plasma, with twelve urinary metabolites being detected in the study [54, 81]. QUE diglucuronides contributed to most of the QUE metabolites in the urine, while other substantial metabolites include QUE 3′-glucuronide, isorhamnetin-3-glucuronide, and QUE glucuronide sulfates (**Table 2**) [81]. Mullen et al. stipulated that the most probable fate of the majority of non-excreted metabolites is the conversion to various phenolic acid compounds, most likely 3-hydroxyphenylpropionic acid, 3,4-dihydroxyphenylpropionic acid, and 3-methoxy-4-hydroyphenylpropionic acid, which were not measured in the study [81, 82].

There is a considerably large interindividual variation in QUE excretion. Ferry et al. examined 51 cancer patients in a phase I clinical trial of quercetin, in which the subjects were administered by a short IV infusion at doses of 60 to 2000 mg/m^2 of QUE aglycone. They reported the proportion of the excreted QUE in urine over 24 hours ranged between 0.03–7.6%, implying variability between subjects [48]. Various studies also examined the t½ of QUE with variable conclusions. Some studies examining the half-life of orally administered QUE reported a t½ ranging between 11

Metabolites	24-h Urinary Excretion	
	Mean (nmol)	% of total excreted metabolites
QUE Diglucuronide	2223	21%
QUE-3'-Glu	1845	18%
I-3-Glu	1739	17%
QUE Glu Sulfate	1384	13%
MQUE Diglucuronide	1003	10%
QUE-3-Glu	912	9%
I-4′-Glu	700	7%
QUE Glucoside Sulfate	392	4%
QUE Glu Glucoside	163	2%
Total	10,361	100%

QUE, quercetin; Glu, glucuronide; nmol, nanomole; and MQUE, methylquercetin.

Table 2.
Mean and percentage of quercetin metabolites detected in urine at 24 hours, adapted and modified from Mullen et al. [81].

and 28 hours [27, 81–83]. On the other hand, two studies evaluating the pharmacokinetics of IV QUE aglycone concluded the t½ of 0.7–2.4 hours [48, 84]. There was consistently significant clearance at 24 hours, and the elimination of QUE was completed within 48 hours of ingestion. Renal clearance after ingestion of 100–200 mg of various QUE glycosides was estimated to be 0.7 L/h. In contrast, Moon et al. reported that renal clearance of 500 mg of QUE aglycone after ingestion was significantly higher, approximately 3.5×10^4 L/h [3, 29]. The urinary recovery, the proportion of excreted drugs through the urine unchanged, was estimated to be $1.0 \pm 0.8\%$.

Tubular cells, in addition to their efflux and influx properties across different membranes, also act as metabolic machinery for QUE in the kidney. The biochemical conversion includes various deconjugation processes of various metabolites to yield aglycones (e.g., deglucuronidation by β-glucuronidase), followed by instantaneous glycosylation by β-glycosyltransferase, sulfation by SULTs, glucuronidation by UGTs, and methylation by COMTs [54].

8. Conclusion and future perspective

QUE is abundant in the human diet and has potential benefits for human health. Dietary supplement of QUE is available over the counter as a free form of QUE aglycone. In contrast to naturally occurring QUE contained in fruits and vegetables, the intended daily doses of QUE supplements are considerably higher, ranging from 500 to 1000 mg. The majority of the absorption of QUE occurs in the small intestine, and insignificant portions are absorbed in both in the stomach and colon. In the intestine, the aglycone is primarily absorbed *via* a passive diffusion mechanism and secondarily *via* OATP, whereas the glucosides are achieved *via* specific transporters. Following the absorption in the enterocytes, QUE undergoes extensive metabolism in the liver, resulting in sulfated, glucuronidated, and/or methylated metabolites.

Based on the pharmacokinetics profile, QUE has poor oral bioavailability after a single oral dose. The bioavailability of QUE is associated with various factors, including its solubility and coadministration of other dietary components. Hence, its formulation to improve absorption rates and reach its optimum plasma level should be considered important. The effects of coadministration of other dietary supplements, such as vitamin C and bromelain, should further be investigated in future clinical studies. It must also be taken into consideration that QUE and its metabolites may interfere with the pharmacokinetics of some drugs by variably interacting with various CYP enzymes; they also inhibit multispecific OATP, ABC, and transporters at micromolar levels.

Funding disclosure

This article received no specific grant from any funding agency in the public, commercial, or not-for-profit sectors.

Conflict of interests

The authors declare no conflict of interest.

Author details

Muhammad Candragupta Jihwaprani[1], Wahyu Choirur Rizky[1] and Mazhar Mushtaq[2]*

1 College of Medicine, Sulaiman Al Rajhi University, Al-Qassim, Kingdom of Saudi Arabia

2 Department of Basic Medical Sciences, Sulaiman Al Rajhi University, Al-Qassim, Kingdom of Saudi Arabia

*Address all correspondence to: dr_hcg@yahoo.com

References

[1] Shen N, Wang T, Gan Q, Liu S, Wang L, Jin B. Plant flavonoids: Classification, distribution, biosynthesis, and antioxidant activity. Food Chemistry. 2022;**383**:132531. DOI: 10.1016/j.foodchem.2022.132531

[2] Michala A-S, Pritsa A. Quercetin: A molecule of great biochemical and clinical value and its beneficial effect on diabetes and cancer. Diseases. 2022;**10**: 37. DOI: 10.3390/diseases10030037

[3] Moon YJ, Wang L, DiCenzo R, Morris ME. Quercetin pharmacokinetics in humans. Biopharmaceutics & Drug Disposition. 2008;**29**:205-217. DOI: 10.1002/bdd.605

[4] Rasouli H, Farzaei MH, Khodarahmi R. Polyphenols and their benefits: A review. International Journal of Food Properties. 2017;**20**:1-42. DOI: 10.1080/10942912.2017.1354017

[5] Rizky WC, Jihwaprani MC, Mushtaq M. Protective mechanism of quercetin and its derivatives in viral-induced respiratory illnesses. Egyptian Journal of Bronchology. 2022;**16**:58. DOI: 10.1186/s43168-022-00162-6

[6] Rizky WC, Jihwaprani MC, Mushtaq M. Protective mechanisms of quercetin in various lung-induced injuries. SCIREA Journal of Clinical Medicine. 2022;**7**:206-223. DOI: 10.54647/cm32844

[7] Rizky WC, Jihwaprani MC, Ansori ANM, Mushtaq M. The pharmacological mechanism of quercetin as adjuvant therapy of COVID-19. Life Research. 2022;**5**:17. DOI: 10.53388/life2022-0205-302

[8] Hatahet T, Morille M, Hommoss A, Dorandeu C, Müller RH, Bégu S. Dermal quercetin smartCrystals®: Formulation development, antioxidant activity and cellular safety. European Journal of Pharmaceutics and Biopharmaceutics. 2016;**102**:51-63. DOI: 10.1016/j.ejpb.2016.03.004

[9] Wadhwa K, Kadian V, Puri V, Bhardwaj BY, Sharma A, Pahwa R, et al. New insights into quercetin nanoformulations for topical delivery. Phytomedicine Plus. 2022;**2**:100257. DOI: 10.1016/j.phyplu.2022.100257

[10] Kaushik D, O'Fallon K, Clarkson PM, Patrick Dunne C, Conca KR, Michniak-Kohn B. Comparison of quercetin pharmacokinetics following oral supplementation in humans. Journal of Food Science. 2012;**77**:H231-H238. DOI: 10.1111/j.1750-3841.2012.02934.x

[11] PubChem Compound Summary for CID 5280343, Quercetin [Internet]. National Center for Biotechnology Information. 2021. Available from: https://pubchem.ncbi.nlm.nih.gov/compound/Quercetin n.d. [Accessed: December 25, 2021]

[12] Anand David AV, Arulmoli R, Parasuraman S. Overviews of biological importance of quercetin: A bioactive flavonoid. Pharmacognosy Reviews. 2016;**10**:84-89. DOI: 10.4103/0973-7847.194044

[13] El-Saber Batiha G, Magdy Beshbishy A, Ikram M, Mulla ZS, Abd El-Hack ME, Taha AE, et al. The pharmacological activity, biochemical properties, and pharmacokinetics of the major natural polyphenolic flavonoid: Quercetin. Food. 2020;**9**:374. DOI: 10.3390/foods9030374

[14] Andres S, Pevny S, Ziegenhagen R, Bakhiya N, Schäfer B, Hirsch-Ernst KI, et al. Safety aspects of the use of quercetin as a dietary supplement. Molecular Nutrition & Food Research. 2018;**62**:1-15. DOI: 10.1002/mnfr.201700447

[15] Larson AJ, Symons JD, Jalili T. Therapeutic potential of quercetin to decrease blood pressure: Review of efficacy and mechanisms. Advances in Nutrition. 2012;**3**:39-46. DOI: 10.3945/an.111.001271

[16] Anwar E, Soliman M, Darwish S, Lotfy H, Tolba M. Mechanistic similarity of immuno-modulatory and anti-viral effects of chloroquine and quercetin (the naturally occurring flavonoid). Clinical Immunology Research. 2020;**4**:1-6. DOI: 10.33425/2639-8494.1028

[17] Egert S, Wolffram S, Bosy-Westphal A, Boesch-Saadatmandi C, Wagner AE, Frank J, et al. Daily quercetin supplementation dose-dependently increases plasma quercetin concentrations in healthy humans. The Journal of Nutrition. 2008;**138**:1615-1621. DOI: 10.1093/jn/138.9.1615

[18] Ou-yang Z, Cao X, Wei Y, Zhang W-W-Q, Zhao M, Duan J. Pharmacokinetic study of rutin and quercetin in rats after oral administration of total flavones of mulberry leaf extract. Revista Brasileira de Farmacognosia. 2013;**23**:776-782. DOI: 10.1590/S0102-695X2013000500009

[19] Lee J, Mitchell AE. Pharmacokinetics of quercetin absorption from apples and onions in healthy humans. Journal of Agricultural and Food Chemistry. 2012;**60**:3874-3881. DOI: 10.1021/jf3001857

[20] D'Andrea G. Quercetin: A flavonol with multifaceted therapeutic applications? Fitoterapia. 2015;**106**:256-271. DOI: 10.1016/j.fitote.2015.09.018

[21] Kasikci MB, Bagdatlioglu N. Bioavailability of quercetin. Current Research in Nutrition and Food Science Journal. 2016;**4**:146-151. DOI: 10.12944/CRNFSJ.4

[22] Day AJ, Williamson G. Biomarkers for exposure to dietary flavonoids: A review of the current evidence for identification of quercetin glycosides in plasma. The British Journal of Nutrition. 2001;**86**:S105-S110. DOI: 10.1079/bjn2001342

[23] Arts IC, Sesink AL, Faassen-Peters M, Hollman PC. The type of sugar moiety is a major determinant of the small intestinal uptake and subsequent biliary excretion of dietary quercetin glycosides. The British Journal of Nutrition. 2004;**91**:841-847. DOI: 10.1079/BJN20041123

[24] Németh K, Plumb GW, Berrin J-G, Juge N, Jacob R, Naim HY, et al. Deglycosylation by small intestinal epithelial cell β-glucosidases is a critical step in the absorption and metabolism of dietary flavonoid glycosides in humans. European Journal of Nutrition. 2003;**42**:29-42. DOI: 10.1007/s00394-003-0397-3

[25] Del Rio D, Rodriguez-Mateos A, Spencer JP, Tognolini M, Borges G, Crozier A. Dietary (poly) phenolics in human health: Structures, bioavailability, and evidence of protective effects against chronic diseases. Antioxidants & Redox Signaling. 2013;**18**:1818-1892. DOI: 10.1089/ars.2012.4581

[26] Almeida AF, Borge GIA, Piskula M, Tudose A, Tudoreanu L, Valentová K, et al. Bioavailability of quercetin in humans with a focus on interindividual variation. Comprehensive Reviews in

Food Science and Food Safety. 2018;**17**:714-731. DOI: 10.1111/1541-4337.12342

[27] Cermak R, Breves G, Lüpke M, Wolffram S. In vitro degradation of the flavonol quercetin and of quercetin glycosides in the porcine hindgut. Archives of Animal Nutrition. 2006;**60**: 180-189. DOI: 10.1080/17450390500467695

[28] Zhang H, Hassan YI, Liu R, Mats L, Yang C, Liu C, et al. Molecular mechanisms underlying the absorption of aglycone and glycosidic flavonoids in a Caco-2 BBe1 cell model. ACS Omega. 2020;5:10782-10793. DOI: 10.1021/acsomega.0c00379

[29] Graefe EU, Wittig J, Mueller S, Riethling A, Uehleke B, Drewelow B, et al. Pharmacokinetics and bioavailability of quercetin glycosides in humans. Journal of Clinical Pharmacology. 2001;**41**:492-499. DOI: 10.1177/00912700122010366

[30] Erlund I, Kosonen T, Alfthan G, Mäenpää J, Perttunen K, Kenraali J, et al. Pharmacokinetics of quercetin from quercetin aglycone and rutin in healthy volunteers. European Journal of Clinical Pharmacology. 2000;**56**: 545-553. DOI: 10.1007/s002280000197

[31] Li S, Liu J, Li Z, Wang L, Gao W, Zhang Z, et al. Sodium-dependent glucose transporter 1 and glucose transporter 2 mediate intestinal transport of quercetin in Caco-2 cells. Food & Nutrition Research. 2020;**64**. DOI: 10.29219/fnr.v64.3745

[32] Chabane MN, Al AA, Peluso J, Muller CD, Ubeaud-Séquier G. Quercetin and naringenin transport across human intestinal Caco-2 cells. The Journal of Pharmacy and Pharmacology. 2009;**61**:1473-1483. DOI: 10.1211/jpp/61.11.0006

[33] Kaşıkcı M, Bağdatlıoğlu N. Bioavailability of quercetin. Current Research in Nutrition and Food Science Journal. 2016;**4**:146-151. DOI: 10.12944/CRNFSJ.4.Special-Issue-October.20

[34] Guo Y, Bruno RS. Endogenous and exogenous mediators of quercetin bioavailability. The Journal of Nutritional Biochemistry. 2015;**26**: 201-210. DOI: 10.1016/j.jnutbio.2014.10.008

[35] Hollman P, de Vries J, van Leeuwen S, Mengelers M, Katan M. Absorption of dietary quercetin glycosides and quercetin in healthy ileostomy volunteers. The American Journal of Clinical Nutrition. 1995;**62**: 1276-1282. DOI: 10.1093/ajcn/62.6.1276

[36] Egert S, Wolffram S, Schulze B, Langguth P, Hubbermann EM, Schwarz K, et al. Enriched cereal bars are more effective in increasing plasma quercetin compared with quercetin from powder-filled hard capsules. The British Journal of Nutrition. 2012;**107**:539-546. DOI: 10.1017/S0007114511003242

[37] Guo Y, Mah E, Davis CG, Jalili T, Ferruzzi MG, Chun OK, et al. Dietary fat increases quercetin bioavailability in overweight adults. Molecular Nutrition & Food Research. 2013;**57**:896-905. DOI: 10.1002/mnfr.201200619

[38] Phuwamongkolwiwat P, Suzuki T, Hira T, Hara H. Fructooligosaccharide augments benefits of quercetin-3-O-β-glucoside on insulin sensitivity and plasma total cholesterol with promotion of flavonoid absorption in sucrose-fed rats. European Journal of Nutrition. 2014;**53**:457-468. DOI: 10.1007/s00394-013-0546-2

[39] Singh S, Jamwal S, Kumar P. Neuroprotective potential of quercetin in combination with piperine

against 1-methyl-4-phenyl-1,2,3,6-tetrahydropyridine-induced neurotoxicity. Neural Regeneration Research. 2017;**12**:1137-1144. DOI: 10.4103/1673-5374.211194

[40] Li H, Li M, Fu J, Ao H, Wang W, Wang X. Enhancement of oral bioavailability of quercetin by metabolic inhibitory nanosuspensions compared to conventional nanosuspensions. Drug Delivery. 2021;**28**:1226-1236. DOI: 10.1080/10717544.2021.1927244

[41] Sequeira IR. Higher doses of ascorbic acid may have the potential to promote nutrient delivery via intestinal paracellular absorption. World Journal of Gastroenterology. 2021;**27**:6750-6756. DOI: 10.3748/wjg.v27.i40.6750

[42] Önal H, Arslan B, Üçüncü Ergun N, Topuz Ş, Yilmaz Semerci S, Kurnaz ME, et al. Treatment of COVID-19 patients with quercetin: A prospective, single center, randomized, controlled trial Turkish Journal of Biology. 2021;**45**: 518-529. DOI: 10.3906/biy-2104-16

[43] Lesjak M, Balesaria S, Skinner V, Debnam ES, Srai SKS. Quercetin inhibits intestinal non-haem iron absorption by regulating iron metabolism genes in the tissues. European Journal of Nutrition. 2019;**58**:743-753. DOI: 10.1007/s00394-018-1680-7

[44] Price G, Patel DA. Drug Bioavailability. Treasure Island: StatPearls Publishing; 2023

[45] Jin F, Nieman DC, Shanely RA, Knab AM, Austin MD, Sha W. The variable plasma quercetin response to 12-week quercetin supplementation in humans. European Journal of Clinical Nutrition. 2010;**64**:692-697. DOI: 10.1038/ejcn.2010.91

[46] Hai Y, Zhang Y, Liang Y, Ma X, Qi X, Xiao J, et al. Advance on the absorption, metabolism, and efficacy exertion of quercetin and its important derivatives absorption, metabolism and function of quercetin. Food Frontiers. 2020;**1**:420-434. DOI: 10.1002/fft2.50

[47] Guo Y, Mah E, Bruno RS. Quercetin bioavailability is associated with inadequate plasma vitamin C status and greater plasma endotoxin in adults. Nutrition. 2014;**30**:1279-1286. DOI: 10.1016/j.nut.2014.03.032

[48] Ferry DR, Smith A, Malkhandi J, Fyfe DW, deTakats PG, Anderson D, et al. Phase I clinical trial of the flavonoid quercetin: Pharmacokinetics and evidence for *in vivo* tyrosine kinase inhibition. Clinical Cancer Research. 1996;**2**:659-668

[49] Chalet C, Rubbens J, Tack J, Duchateau GS, Augustijns P. Intestinal disposition of quercetin and its phase-II metabolites after oral administration in healthy volunteers. The Journal of Pharmacy and Pharmacology. 2018;**70**: 1002-1008. DOI: 10.1111/jphp.12929

[50] Chandra P, Brouwer KL. The complexities of hepatic drug transport: Current knowledge and emerging concepts. Pharmaceutical Research. 2004;**21**:719-735. DOI: 10.1023/b:pham.0000026420.79421.8f

[51] Glaeser H, Bujok K, Schmidt I, Fromm MF, Mandery K. Organic anion transporting polypeptides and organic cation transporter 1 contribute to the cellular uptake of the flavonoid quercetin. Naunyn-Schmiedeberg's Archives of Pharmacology. 2014;**387**: 883-891. DOI: 10.1007/s00210-014-1000-6

[52] Kalliokoski A, Niemi M. Impact of OATP transporters on pharmacokinetics. British Journal of Pharmacology. 2009;

158:693-705. DOI: 10.1111/j.1476-5381.2009.00430.x

[53] Ziberna L, Fornasaro S, Čvorović J, Tramer F, Passamonti S. Bioavailability of flavonoids: The role of cell membrane transporters. In: Polyphenols: Mechanisms of Action in Human Health and Disease. Elsevier; 2014. pp. 489-511. DOI: 10.1016/B978-0-12-398456-2.00037-2

[54] Muñoz-Reyes D, Morales AI, Prieto M. Transit and metabolic pathways of quercetin in tubular cells: Involvement of its antioxidant properties in the kidney. Antioxidants. 2021;**10**:909. DOI: 10.3390/antiox10060909

[55] Yin H, Ma J, Han J, Li M, Shang J. Pharmacokinetic comparison of quercetin, isoquercitrin, and quercetin-3-O-β-D-glucuronide in rats by HPLC-MS. PeerJ. 2019;7:e6665. DOI: 10.7717/peerj.6665

[56] Yang L-L, Xiao N, Li X-W, Fan Y, Alolga RN, Sun X-Y, et al. Pharmacokinetic comparison between quercetin and quercetin 3-O-β-glucuronide in rats by UHPLC-MS/MS. Scientific Reports. 2016;**6**:35460. DOI: 10.1038/srep35460

[57] Zhou S, Chan E, Duan W, Huang M, Chen Y-Z. Drug bioactivation covalent binding to target proteins and toxicity relevance. Drug Metabolism Reviews. 2005;**37**:41-213. DOI: 10.1081/dmr-200028812

[58] Rastogi H, Jana S. Evaluation of inhibitory effects of caffeic acid and quercetin on human liver cytochrome p450 activities. Phytotherapy Research. 2014;**28**:1873-1878. DOI: 10.1002/ptr.5220

[59] Williamson G, Day A, Plumb G, Couteau D. Human metabolic pathways of dietary flavonoids and cinnamates. Biochemical Society Transactions. 2000; **28**:16-22. DOI: 10.1042/bst0280016

[60] Zhang L, Zuo Z, Lin G. Intestinal and hepatic glucuronidation of flavonoids. Molecular Pharmaceutics. 2007;**4**:833-845. DOI: 10.1021/mp700077z

[61] O'Leary KA, Day AJ, Needs PW, Mellon FA, O'Brien NM, Williamson G. Metabolism of quercetin-7-and quercetin-3-glucuronides by an *in vitro* hepatic model: The role of human β-glucuronidase, sulfotransferase, catechol-O-methyltransferase and multi-resistant protein 2 (MRP2) in flavonoid metabolism. Biochemical Pharmacology. 2003;**65**:479-491. DOI: 10.1016/s0006-2952(02)01510-1

[62] Liu Y, Liu Y, Dai Y, Xun L, Hu M. Enteric disposition and recycling of flavonoids and ginkgo flavonoids. Journal of Alternative and Complementary Medicine. 2003;**9**: 631-640. DOI: 10.1089/107555303322524481

[63] Lotito SB, Zhang W-J, Yang CS, Crozier A, Frei B. Metabolic conversion of dietary flavonoids alters their anti-inflammatory and antioxidant properties. Free Radical Biology & Medicine. 2011;**51**:454-463. DOI: 10.1016/j.freeradbiomed.2011.04.032

[64] Chen J, Wang S, Jia X, Bajimaya S, Lin H, Tam VH, et al. Disposition of flavonoids via recycling: Comparison of intestinal versus hepatic disposition. Drug Metabolism and Disposition. 2005; **33**:1777-1784. DOI: 10.1124/dmd.105.003673

[65] Vázquez-Flores L, Casas-Grajales S, Hernández-Aquino E, Vargas-Pozada E,

Muriel P. Antioxidant, antiinflammatory, and antifibrotic properties of quercetin in the liver. In: Liver Pathophysiology. Elsevier; 2017. pp. 653-674. DOI: 10.1016/B978-0-12-804274-8.00047-3

[66] Janisch KM, Williamson G, Needs P, Plumb GW. Properties of quercetin conjugates: Modulation of LDL oxidation and binding to human serum albumin. Free Radical Research. 2004;**38**: 877-884. DOI: 10.1080/10715760410001728415

[67] Kawai Y, Nishikawa T, Shiba Y, Saito S, Murota K, Shibata N, et al. Macrophage as a target of quercetin glucuronides in human atherosclerotic arteries: Implication in the anti-atherosclerotic mechanism of dietary flavonoids. The Journal of Biological Chemistry. 2008;**283**:9424-9434. DOI: 10.1074/jbc.M706571200

[68] Nguyen MA, Staubach P, Wolffram S, Langguth P. Effect of single-dose and short-term administration of quercetin on the pharmacokinetics of talinolol in humans–implications for the evaluation of transporter-mediated flavonoid–drug interactions. European Journal of Pharmaceutical Sciences. 2014;**61**:54-60. DOI: 10.1016/j.ejps.2014.01.003

[69] Poór M, Boda G, Needs PW, Kroon PA, Lemli B, Bencsik T. Interaction of quercetin and its metabolites with warfarin: Displacement of warfarin from serum albumin and inhibition of CYP2C9 enzyme. Biomedicine & Pharmacotherapy. 2017; **88**:574-581. DOI: 10.1016/j.biopha.2017.01.092

[70] Nguyen MA, Staubach P, Wolffram S, Langguth P. The influence of single-dose and short-term administration of quercetin on the pharmacokinetics of midazolam in humans. Journal of Pharmaceutical Sciences. 2015;**104**:3199-3207. DOI: 10.1002/jps.24500

[71] Kim KA, Park PW, Park JY. Short-term effect of quercetin on the pharmacokinetics of fexofenadine, a substrate of P-glycoprotein, in healthy volunteers. European Journal of Clinical Pharmacology. 2009;**65**:609-614. DOI: 10.1007/s00228-009-0627-6

[72] Singh BN, Upreti DK, Singh BR, Pandey G, Verma S, Roy S, et al. Quercetin sensitizes fluconazole-resistant *Candida albicans* to induce apoptotic cell death by modulating quorum sensing. Antimicrobial Agents and Chemotherapy. 2015;**59**:2153-2168. DOI: 10.1128/AAC.03599-14

[73] Gao M, Wang H, Zhu L. Quercetin assists fluconazole to inhibit biofilm formations of fluconazole-resistant *Candida albicans in vitro* and *in vivo* antifungal managements of vulvovaginal candidiasis. Cellular Physiology and Biochemistry. 2016;**40**:727-742. DOI: 10.1159/000453134

[74] Vipin C, Saptami K, Fida F, Mujeeburahiman M, Rao SS, Arun AB, et al. Potential synergistic activity of quercetin with antibiotics against multidrug-resistant clinical strains of *Pseudomonas aeruginosa*. PLoS One. 2020; **15**:e0241304. DOI: 10.1371/journal.pone.0241304

[75] Siriwong S, Teethaisong Y, Thumanu K, Dunkhunthod B, Eumkeb G. The synergy and mode of action of quercetin plus amoxicillin against amoxicillin-resistant *Staphylococcus epidermidis*. BMC Pharmacology and Toxicology. 2016;**17**:1-14. DOI: 10.1186/s40360-016-0083-8

[76] Qu S, Dai C, Shen Z, Tang Q, Wang H, Zhai B, et al. Mechanism of synergy between tetracycline and quercetin against antibiotic resistant *Escherichia coli*. Frontiers in Microbiology. 2019;**10**:2536. DOI: 10.3389/fmicb.2019.02536

[77] Dagher O, Mury P, Thorin-Trescases N, Noly PE, Thorin E, Carrier M. Therapeutic potential of quercetin to alleviate endothelial dysfunction in age-related cardiovascular diseases. Frontiers in Cardiovascular Medicine. 2021;**8**:220. DOI: 10.3389/fcvm.2021.658400

[78] Chen Y-Q, Chen H Y, Tang Q Q, Li Y-F, Liu X-S, Lu F-H, et al. Protective effect of quercetin on kidney diseases: From chemistry to herbal medicines. Frontiers in Pharmacology. 2022;**13**: 968226. DOI: 10.3389/fphar.2022.968226

[79] Wong CC, Botting NP, Orfila C, Al-Maharik N, Williamson G. Flavonoid conjugates interact with organic anion transporters (OATs) and attenuate cytotoxicity of adefovir mediated by organic anion transporter 1 (OAT1/SLC22A6). Biochemical Pharmacology. 2011;**81**:942-949. DOI: 10.1016/j.bcp.2011.01.004

[80] Wong CC, Akiyama Y, Abe T, Lippiat JD, Orfila C, Williamson G. Carrier-mediated transport of quercetin conjugates: Involvement of organic anion transporters and organic anion transporting polypeptides. Biochemical Pharmacology. 2012;**84**:564-570. DOI: 10.1016/j.bcp.2012.05.011

[81] Mullen W, Edwards CA, Crozier A. Absorption, excretion and metabolite profiling of methyl-, glucuronyl-, glucosyl- and sulpho-conjugates of quercetin in human plasma and urine after ingestion of onions. The British Journal of Nutrition. 2006;**96**:107. DOI: 10.1079/BJN20061809

[82] Olthof MR, Hollman PCH, Buijsman MNCP, JMM VA, Katan MB. Chlorogenic acid, Quercetin-3-rutinoside and black tea phenols are extensively metabolized in humans. The Journal of Nutrition. 2003;**133**: 1806-1814. DOI: 10.1093/jn/133.6.1806

[83] Manach C, Williamson G, Morand C, Scalbert A, Rémésy C. Bioavailability and bioefficacy of polyphenols in humans. I. Review of 97 bioavailability studies. The American Journal of Clinical Nutrition. 2005;**81**: 230S-242S. DOI: 10.1093/ajcn/81.1.230S

[84] Gugler R, Leschik M, Dengler HJ. Disposition of quercetin in man after single oral and intravenous doses. European Journal of Clinical Pharmacology. 1975;**9**:229-234. DOI: 10.1007/BF00614022

Section 3

Molecular Pathways

Chapter 4

Leveraging a Hypothesis-Generating Transcriptomics Approach to Elucidate Molecular Pathways that Contribute to the Biologic Effects of Quercetin in the Liver

Nhung Au and Brendan D. Stamper

Abstract

Quercetin is a relatively ubiquitous natural product with reported antioxidant, anti-inflammatory, antiviral, and anticarcinogenic properties. Using a bioinformatics approach, differential gene expression analysis was utilized to evaluate quercetin's potential to protect and promote hepatocellular health through mining of the Gene Expression Omnibus (GEO) and subsequent analysis using the Database for Annotation and Visualization and Integrated Discovery (DAVID). The publicly available microarray datasets GSE4259 and GSE72081 were analyzed to compare the effect of quercetin on two different liver-based model systems to generate a robust set of differentially expressed genes impacted by quercetin exposure. Results from these analyses identified differentially expressed genes related to calcium signaling and signal transduction pathways to be the most significantly altered. A comprehensive literature review following the transcriptome analysis revealed that quercetin-induced gene expression changes in cell membrane receptors (specifically, voltage gated calcium channels NS integrins) share a common direct signaling pathway through extracellular signal-regulated kinase (ERK). Thus, the results from this bioinformatics study identified potential biomarkers related to quercetin's effects on hepatocellular health. Based on quercetin's ubiquitous use and good safety profile, future laboratory studies can be directed at validating the observed transcriptional changes on protein expression and the likelihood for hepatoprotection.

Keywords: quercetin, extracellular signal-regulated kinase, gene expression, signal transduction, bioinformatics

1. Introduction

Natural products have been used throughout human history for various purposes due to the ability of their chemical constituents to exhibit various pharmacologic activities in humans, many of which have proven to be beneficial for treating and preventing a broad spectrum of diseases [1]. To this day, natural products continue to be an excellent reservoir for lead compound development in drug discovery research. For example, structure-based drug design with a novel nature-based structural skeleton has been used to improve the pharmacodynamic and pharmacokinetic characteristics of a natural product [2]. While there is little doubt, natural products represent a rich diversity of potential drug candidates, challenges such as a lack of well-designed candidate selecting strategies may hinder progress and limit expansion in the field. Despite these challenges, there is continued interest in natural products research from both the public and within the scientific community, which has led to the development of new technologies and approaches to increase the likelihood of a successful line of inquiry into the potential of a given natural product as a future drug candidate [3, 4].

Quercetin (3,3′,4′,5,7-pentahydroxy-flavone) is a polyphenolic compound derived from plant pigments and is a natural product found in many different plant species such as grapes, berries, cherries, apples, citrus fruit, kale, and black tea [5]. In plants, quercetin mainly presents as quercetin glycosides, and exhibits different therapeutic properties, which have been shown to have beneficial effects in treating various disease states with pathophysiologies related to oxidative stress, inflammation, and the immune system [6, 7]. While more work needs to be done to elucidate the precise mechanisms underlying how quercetin exerts its effects on cells, efforts must also be taken to make quercetin more "druggable". From a pharmacokinetic perspective, quercetin struggles with low bioavailability and extensive metabolism [5]. Both enterohepatic and enteric recycling are thought to contribute to its poor systemic bioavailability, similar to many other quercetin-like polyphenolic structures [8]. Based on the fact that the liver has the potential to be exposed to greater quercetin concentrations compared to other systemic organs, quercetin may possess utility in treating hepatic ailments and promoting hepatocellular health.

Through investigations into how quercetin promotes hepatocellular health, researchers have identified a variety of hepatic signaling pathways that are altered following quercetin exposure, such as those associated with antioxidant defense, MAPKs (mitogen-activated protein kinases), inflammation, apoptosis, autophagy, insulin, AMPK (AMP-activated protein kinase), β-Catenin signaling, and antiviral activity [9]. For example, quercetin has been shown to elicit a dose-dependent protective effect against ethanol-induced oxidative stress in human hepatocytes [10]. This protection was found to occur through MAPK-mediated nuclear Nrf2 (nuclear factor erythroid 2-related factor 2) translocation and subsequent induction of heme oxygenase-1 (HO-1). Chronic inflammation also plays a role in various liver disorders such as non-alcoholic fatty liver disease (NAFLD) and hepatocellular carcinoma. Research suggests quercetin is capable of reducing liver inflammation and fibrosis through inhibition of macrophage infiltration and hepatic stellate cell activation [11]. In addition, quercetin's ability to inhibit NF-kB (nuclear factor-kappa B) signaling may also be a contributing factor to its anti-inflammatory effects [12]. Quercetin's significant effects on NF-kB signaling, MAPK pathways, and the antioxidant response system have buoyed its popularity as a promising and druggable natural product since these facotrs are associated with important cellular processes such as proliferation, differentiation, stress response, inflammation, and apoptosis.

Over the past thirty years, numerous studies have identified other pathways affected by quercetin exposure, many of which crosstalk with NF-kB, MAPK and Nrf2. In general, quercetin has been shown to modulate PI3K/Akt (phosphatidylinositol 3-kinase/protein kinase B) signaling, mTOR (mammalian target of rapamycin), and AMPK (AMP-activated protein kinase), which in turn influence cellular effects related to autophagy [13, 14], insulin resistance [15], lipophagy [16], and apoptosis [17]. Interestingly, quercetin may also have utility as a chemosensitizer [18]. This study found quercetin capable of reversing multidrug resistance in hepatocellular carcinoma through the FZD7 (Frizzled-7)/β-catenin signaling, a pathway involved in liver development and regeneration. Taken together, the growing body of literature covering quercetin-mediated effects on the liver has shown that quercetin alters numerous hepatic signaling pathways across a diverse range of model systems and over a wide variety of treatment conditions; thus, implicating a complex, multifaceted, and "dirty" mechanism by which quercetin is able to promote hepatocellular health.

In clinical trials, quercetin has shown promise in the treatment and protection of the liver from injury and damage in various disease states. In patients with beta-thalassemia, quercetin was coadministered with deferoxamine and demonstrated an ability to lower alanine aminotransferase (ALT) levels and positively affect liver function [19]. However, in a separate trial looking at patients with non-alcoholic fatty liver disease (NAFLD), quercetin showed no significant effect on hepatic function biomarkers and liver enzymes [20]. Quercetin's potential for hepatoprotection has also been investigated for the treatment of chronic hepatitis C infection [21]. While no changes in ALT levels were observed in this study, a decrease in viral load for 8 of 30 patients taking quercetin was observed. It is also worth noting that across all of these studies, quercetin displayed a good safety profile [22]. Yet, while some of this clinical data shows promise and supplementary compelling evidence from animal studies is encouraging, progress has been slow and the need for more extensive clinical trials is needed to validate the potential positive effects of quercetin for patients with various liver conditions [23, 24].

One popular and common method for assessing how a compound might affect biology is transcriptome analysis. Microarray technology has existed for approximately forty years and has provided researchers and clinicians an opportunity to better understand transcriptomic changes at a global level [25]. Observed gene expression profile changes that can then be applied to curated biochemical pathways to posit the possible impact on human health and disease. With an ability to analyze the expression levels of thousands of different loci simultaneously, whole transcriptome microarrays offer an attractive high throughput method for researchers to assess a complete picture of gene expression in response to a variety of conditions, such as xenobiotic exposure and disease status. In the early 2000s, the National Center for Biotechnology Information (NCBI) and the European Bioinformatics Institute (EBI) established the Gene Expression Omnibus (GEO) [26] and Array Express [27], respectively to house the plethora of functional genomics data that has been generated over the past half century. A positive component of these two repositories is that the data is publicly available and follows a standardized format for inclusion known as minimum information about a microarray experiment (MIAME) [28]. Free and unfettered access allows objective investigators, unaffiliated with the original researchers who deposited the data, to probe and analyze these datasets independently. In many cases, the original depositors of a microarray experiment are querying their data to ask specific questions. This allows the unaffiliated researcher to compare data against other deposited datasets to leverage multiple studies across multiple species utilizing multiple platforms to elucidate biochemical pathways, identify robust biomarkers, and cultivate potential drug targets. Despite quercetin's relatively ubiquitous use as a natural product, there are surprisingly few microarray studies on its transcriptomic effects in the liver and liver-based models. The following work showcases how bioinformatic data repositories and platforms can be used as tools for hypothesis-generation in the laboratory by using quercetin-induced hepatocellular effects as an exemplar. Two microarray studies (one from Canada and the other from the Netherlands), which investigated the effects of quercetin on two different liver models, were leveraged to identify biomarkers and pathways in the liver that are impacted following quercetin exposure.

2. Methodology

2.1 Transcriptome database mining

The Gene Expression Omnibus (GEO, https://www.ncbi.nlm.nih.gov/geo/) is an international public repository that archives and freely distributes microarray, next-generation sequencing, and other forms of high throughput functional genomics data. In this study, data from two separate studies investigating the effects of quercetin exposure were mined from GEO; specifically, GSE4259 [29], a study in C57BL/6 mice and GSE72081 [30], a study using primary mouse hepatocytes isolated from C57BL/6 mice. Raw gene expression data was downloaded from GEO and imported into Microsoft Excel for single gene analysis. Transcripts from quercetin-treated samples were compared against their respective controls to generate $\log_2$ fold changes, as well as non-log transformed fold changes and p-values (student T-test). The data was then filtered using a threshold ($p < 0.05$) to identify all transcripts with significantly altered expression. Transcripts from quercetin-treated samples demonstrating a significant change in expression compared to their respective controls were compiled and designated as differentially expressed gene sets.

2.2 Bioinformatic pathway analysis

Differentially expressed gene sets were then submitted for analysis using the Database for Annotation, Visualization and Integrated Discovery (DAVID, https://david.ncifcrf.gov/) platform. DAVID is a web-accessible tool that provides a comprehensive set of functional annotation tools for investigators to understand biological meaning behind large gene lists. In this study, DAVID was used to conduct pathway analysis in order to identify Kyoto Encyclopedia of Genes and Genomes pathways (KEGG; https://www.genome.jp/kegg/pathway.html) impacted by the GEO-identified differentially expressed gene sets. Only pathways that met a significance cutoff of (p-value <0.05) using DAVID were reported. Of note, individual genes identified across multiple pathways were grouped together and dubbed the core gene battery.

3. Hepatic transcriptome changes associated with quercetin exposure

3.1 Conceptualization of a core gene battery

Natural products serve as a great resource for identifying potential drug candidates, and while many possess attractive pharmacodynamic activities, many are also limited by their pharmacokinetics. Quercetin is no exception to this. Like a double-edged sword, quercetin has attractive pharmacodynamic properties such as antioxidant, anti-inflammatory, and immune-supportive properties; however, quercetin also

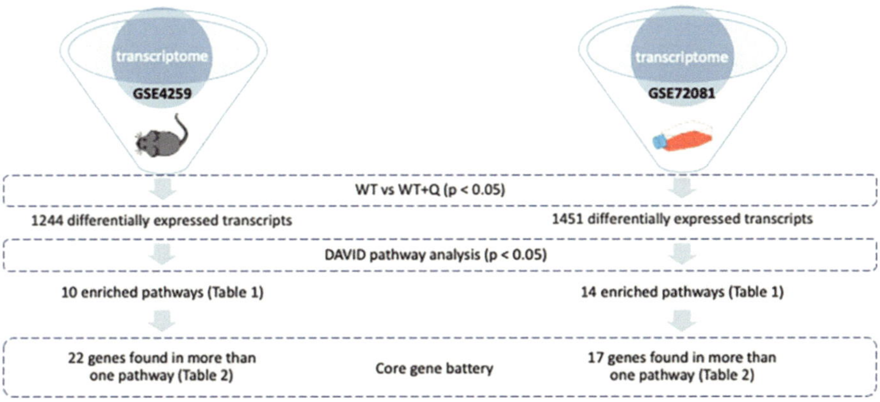

Figure 1.
Workflow for the identification of differentially expressed genes and enriched biologic pathways across two quercetin studies.

has poor bioavailability due to extensive metabolism, enterohepatic recirculation, and enteric recycling. Despite the challenges associated with delivery and retention of quercetin in the human body, a growing body of literature continues to suggest that quercetin is capable of offering positive health effects, including hepatoprotection.

In this study, two independent microarray experiments were leveraged to investigate the effects of quercetin on two different liver models in hopes of identifying robust biomarkers and mechanisms associated with quercetin exposure. One study treated male C57BL/6 mice with 7 mg quercetin per mouse for 24 h (GSE4259) [29], whereas the other utilized primary mouse hepatocytes treated with 200uM quercetin for 24 h (GSE72081) [30]. Filtering the gene sets from these two studies based on significance ($p < 0.05$) resulted in the identification of over 1000 transcripts differentially regulated by quercetin treatment in each study (**Figure 1**). For the in vivo study, the 1244 differentially expressed genes were associated with 10 KEGG pathways, whereas 14 KEGG pathways were found to be enriched from the set of 1451 differentially expressed genes from the in vitro study (**Table 1**).

Interestingly, pathway analysis using DAVID identified no overlapping pathways between the two studies (**Table 1**). In an attempt to uncover commonalities between the two quercetin studies, a unique approach of identifying genes found in multiple enriched KEGG pathways was employed. Thus, individual genes identified across multiple pathways were grouped together and dubbed the core gene battery (**Table 2**). Similar to the enriched pathways identified by DAVID, no genes within the core gene batteries were shared between the two quercetin studies. There are certainly limitations based on the experimental design of this bioinformatic study that could explain why identical transcripts were not found between the two independent studies. First, retrospective bioinformatic surveys like this are always limited by the data that is accessible. In this case, only two studies investigating the effects of quercetin on hepatic murine gene expression were available (one using C57BL/6 mice that was published in 2006 and the other using primary mouse hepatocytes from C57BL/6 mice that was published in 2016). This limited data posed two significant challenges. First, the two independent studies were separated by 10 years, and significant improvements in how gene expression is quantified were made during this time frame. Second, while the same mouse strain was used in both studies, one performed whole liver analysis, while the other assayed isolated primary hepatocytes. Differences in how the tissue was prepared for

Enriched KEGG pathways from GSE4259 [29]	Enriched KEGG pathways from GSE72081 [30]
• Inositol phosphate metabolism • Dilated cardiomyopathy • Cysteine and methionine metabolism • Hypertrophic cardiomyopathy (HCM) • Arrhythmogenic right ventricular cardiomyopathy (ARVC) • Phosphatidylinositol signaling system • Primary bile acid biosynthesis • Amino sugar and nucleotide sugar metabolism • Calcium signaling pathway • Cardiac muscle contraction	• NOD-like receptor signaling pathway • Vasopressin-regulated water reabsorption • Insulin signaling pathway • RNA degradation • Protein processing in endoplasmic reticulum • Type II diabetes mellitus • Retrograde endocannabinoid signaling • Lysine degradation • MAPK signaling pathway • Adipocytokine signaling pathway • Adherents junction • Prolactin signaling pathway • cAMP signaling pathway • TNF signaling pathway

Table 1.
Enriched pathways (p-value <0.05) associated with quercetin exposure from two independent studies.

GSE4259 Core gene battery [29]		GSE72081 Core gene battery [30]	
Upregulated	**Downregulated**	**Upregulated**	**Downregulated**
ACTB	CACNB4	CXCL2	AKT1
CACNA2D3	CACNG2	GRIA4	CACNA1D
CACNB3	ITGA3	MAPK10	GRIA3
ITGA4	ITGB5	MKNK1	IRS2
ITGA7	ITPR2	NFKB1	MAPK3
ITGA9	MAT2A	SOCS2	MAPK12
PIK3C2G	MYBPC3	SOCS3	PRKCA
PIP4K2C	PIP5K1B	TGFB1	SORBS1
PIP5K1A	PLCB2		TRAF6
PLCG2	PRKACB		
SLC8A1	TGFB2		

Table 2.
Core gene batteries associated with quercetin exposure from two independent studies.

analysis can impact effects and lead to divergent results. However, despite the fact that there was no overlap in biochemical pathways or specific genes between the two studies, the identification of genes found in multiple enriched pathways through DAVID revealed that many of the genes within the core gene batteries of each study encoded for related isoforms (e.g., transforming growth factor-beta) or different subunits of the same protein (e.g., voltage-gated calcium channels) (**Table 2**).

From an objective big-picture perspective, the results from this pilot bioinformatics study suggest that quercetin alters the transcription of membrane proteins related to calcium signaling and specific signal transduction pathways that appear to consistently funnel through the extracellular signal-regulated kinase (ERK) (**Figure 2**). The three membrane proteins and one cytokine that were identified from the comparative microarray study are voltage-gated calcium channels (VGCCs), sodium-calcium

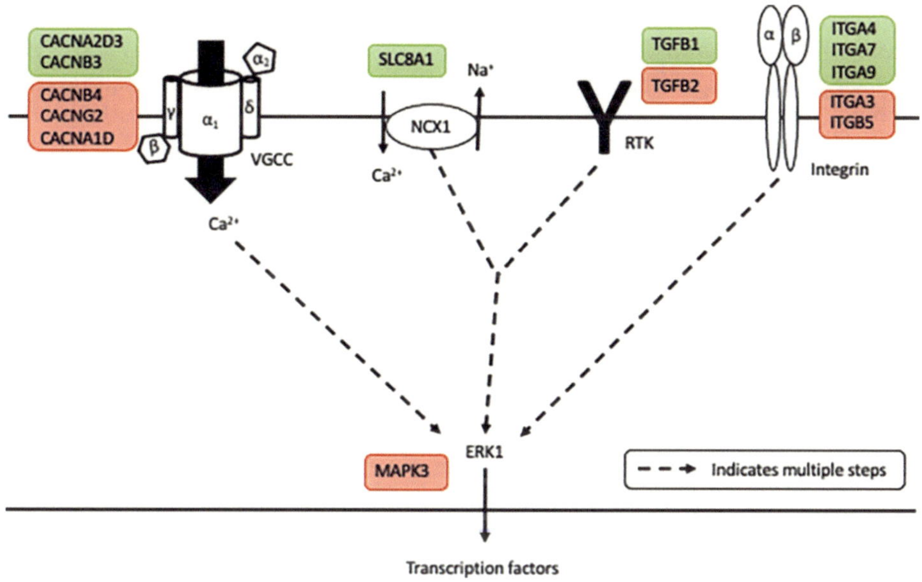

Figure 2.
Signaling pathways that link membrane proteins, calcium signaling, and ERK with genes identified through DAVID analysis. Upregulated transcripts that are part of the core gene battery are boxed in green whereas downregulated transcripts that are part of the core gene battery are boxed in red.

exchange protein (NCX; SLC8A1), integrins, and the cytokine, transforming growth factor beta 2 (TGFB2), whose general roles are all described in what follows.

3.2 Transcripts within the core gene battery and their convergence on ERK

Calcium signaling plays a crucial role in liver function and survival, and is regulated by changes in the intracellular concentration of calcium, which is highly organized by time and space. VGCCs play a major role in the regulation of intracellular calcium, which in turn indirectly activates mitogen-activated protein kinase (MAPK) signal transduction pathways (e.g., ERK 1/2) [31]. VGCCs are composed of multiple subunits (alpha1, alpha2delta, beta, and gamma), where ion conduction occurs through the alpha1 subunit, whereas other subunits play modulatory roles [32, 33]. Like VGCCs, NCX is also capable of regulating intracellular calcium concentration. While expressed mainly in the heart and brain, little is known regarding its role in the liver. Mechanistically, NCX has two working modes: (1) forward-mode, which pumps Ca^{2+} out of the cell in exchange for Na^{+}, and (2) reverse-mode, which brings Ca^{2+} into the cell and pumps Na^{+} out in a concentration-dependent fashion. In reverse-mode, it has been proposed (through a stepwise process) that NCX plays a critical role in downstream ERK1/2 phosphorylation in endothelial cells [34]. It's also worth noting that the inositol 1,4,5-trisphosphate receptor type 2 (ITPR2), which is expressed in the endoplasmic reticulum and serves to release calcium to the mitochondria as part of senescence was identified as part of the downregulated core gene battery as well (**Table 2**) [35]. This further implicates the likelihood that quercetin exposure alters intracellular calcium signaling.

Two additional targets identified by GEO and DAVID as differentially expressed following quercetin treatment were TGFB and integrins, specifically, TGFB1, TGFB2, ITGA3, ITGA4, ITGA7, ITGA9, and ITGB5. While these targets have no direct effect on calcium flux in or out of the cell, they are both capable of initiating signals leading to ERK activation. For example, it is well-established that calcium signals mediate tyrosine kinase receptor signaling, which in turn activates ERK-mediated signaling; thus, providing a possible connection between TGFB and ERK [36–38]. And lastly, integrins, which are heterometric transmembrane cell adhesion proteins, are well-characterized regulators of cell growth, proliferation, migration, signaling, and cytokine activity. Integrins are formed by the pairing integrin alpha and beta subunits. There are 18 different α subunits which are capable of combining with eight different β subunits to form several unique integrin heterodimers [39]. Despite the complexities associated with integrin dimers, strong evidence exists supporting the fact that integrin-mediated signaling regulates ERK phosphorylation and subsequent ERK activity [40].

By altering the gene expression of membrane proteins, and potentially their composition and localization, quercetin has the potential to alter calcium signals and ERK-mediated signaling in the liver. Perhaps it was foreseeable that the three membrane proteins and the one cytokine that were highlighted herein from the comparative microarray study were all capable of modulating ERK activity since it has well established that growth factors and adhesion signals activate ERK. These effects align with quercetin's purported common pharmacodynamic effects that have been reported in the literature related to cell proliferation, regeneration, and adhesion. A confounding factor that was challenging to address was the discovery of simultaneous up and downregulation of specific subunits of VGCCs and integrins. Making sense of how the inconsistent, yet significant, regulation of these subunits relates to a predictive cellular response is challenging because each specific combination of these variable subunits gives rise to a unique mode of activity.

4. Conclusions

In summary, the GEO- and DAVID-based results identified cell signaling pathways that may contribute to how quercetin affects hepatocellular health. These results highlight how bioinformatic approaches can be leveraged to generate meaningful and testable hypotheses in the laboratory. Next steps should involve validation through protein expression and activity studies related to calcium signaling. While the results from this study implicate membrane proteins related to calcium signaling and potentially ERK, additional efforts should be directed at investigating effects stemming from quercetin's structure, such as its chemical ability to redox cycle and its reported ability to serve as both a direct and indirect antioxidant. This would include future cytoprotection studies to investigate quercetin's ability to protect against known toxins capable of causing oxidative stress. While these cell-based assays are necessary for validating quercetin as a useful therapeutic agent for hepatoprotection, the preliminary bioinformatic results from this pilot study have provided intriguing targets and potential mechanisms related to quercetin's future as an agent capable of hepatoprotection. Dissemination of hypothesis-generating results and the identities of these gene targets and mechanistic pathways, provides a record and a springboard on which the scientific community can build.

Conflict of interest

The authors declare no conflict of interest.

Author details

Nhung Au and Brendan D. Stamper*
Pacific University School of Pharmacy, Hillsboro, OR, United States

*Address all correspondence to: stamperb@pacificu.edu

References

[1] Dias DA, Urban S, Roessner U. A historical overview of natural products in drug discovery. Metabolites. 2012;**2**(2):303-336. DOI: 10.3390/metabo2020303

[2] Newman DJ, Cragg GM. Natural products as sources of new drugs from 1981 to 2014. Journal of Natural Products. 2016;**79**(3):629-661. DOI: 10.1021/acs.jnatprod.5b01055

[3] Atanasov AG, Zotchev SB, Dirsch VM, International Natural Product Sciences Taskforce, Supuran CT. Natural products in drug discovery: Advances and opportunities. Nature Reviews. Drug Discovery. 2021;**20**(3):200-216. DOI: 10.1038/s41573-020-00114-z

[4] Najmi A, Javed SA, Al Bratty M, Alhazmi HA. Modern approaches in the discovery and development of plant-based natural products and their analogues as potential therapeutic agents. Molecules. 2022;**27**(2):349. DOI: 10.3390/molecules27020349

[5] Salehi B et al. Therapeutic potential of quercetin: New insights and perspectives for human health. ACS Omega. 2020;**5**(20):11849-11872. DOI: 10.1021/acsomega.0c01818

[6] Li Y et al. Quercetin, inflammation and immunity. Nutrients. 2016;**8**(3):167. DOI: 10.3390/nu8030167

[7] Anand David AV, Arulmoli R, Parasuraman S. Overviews of biological importance of quercetin: A bioactive flavonoid. Pharmacognosy Reviews. 2016;**10**(20):84-89. DOI: 10.4103/0973-7847.194044

[8] Chen J, Lin H, Hu M. Metabolism of flavonoids via enteric recycling: Role of intestinal disposition. The Journal of Pharmacology and Experimental Therapeutics. 2003;**304**(3):1228-1235. DOI: 10.1124/jpet.102.046409

[9] Zhao X et al. Quercetin as a protective agent for liver diseases: A comprehensive descriptive review of the molecular mechanism. Phytotherapy Research. 2021;**35**(9):4727-4747. DOI: 10.1002/ptr.7104

[10] Yao P et al. Quercetin protects human hepatocytes from ethanol-derived oxidative stress by inducing heme oxygenase-1 via the MAPK/Nrf2 pathways. Journal of Hepatology. 2007;**47**(2):253-261. DOI: 10.1016/j.jhep.2007.02.008

[11] Li X et al. The flavonoid quercetin ameliorates liver inflammation and fibrosis by regulating hepatic macrophages activation and polarization in mice. Frontiers in Pharmacology. 2018;**9**:72. DOI: 10.3389/fphar.2018.00072

[12] Granado-Serrano AB, Martín MÁ, Bravo L, Goya L, Ramos S. Quercetin attenuates TNF-induced inflammation in hepatic cells by inhibiting the NF-κB pathway. Nutrition and Cancer. 2012;**64**(4):588-598. DOI: 10.1080/01635581.2012.661513

[13] Wu L et al. Quercetin prevents hepatic fibrosis by inhibiting hepatic stellate cell activation and reducing autophagy via the TGF-β1/Smads and PI3K/Akt pathways. Scientific Reports. 2017;**7**(1):9289. DOI: 10.1038/s41598-017-09673-5

[14] Liu L, Gao C, Yao P, Gong Z. Quercetin alleviates high-fat diet-induced oxidized low-density lipoprotein

accumulation in the liver: Implication for autophagy regulation. BioMed Research International. 2015;**2015**:607531. DOI: 10.1155/2015/607531

[15] Liu H et al. Enhanced alleviation of insulin resistance via the IRS-1/Akt/FOXO1 pathway by combining quercetin and EGCG and involving miR-27a-3p and miR-96-5p. Free Radical Biology & Medicine. 2022;**181**:105-117. DOI: 10.1016/j.freeradbiomed.2022.02.002

[16] Zeng H et al. Quercetin alleviates ethanol-induced liver steatosis associated with improvement of lipophagy. Food and Chemical Toxicology. 2019;**125**:21-28. DOI: 10.1016/j.fct.2018.12.028

[17] Sethi G et al. Apoptotic mechanisms of quercetin in liver cancer: Recent trends and advancements. Pharmaceutics. 2023;**15**(2):712. DOI: 10.3390/pharmaceutics15020712

[18] Chen Z et al. Reversal effect of quercetin on multidrug resistance via FZD7/β-catenin pathway in hepatocellular carcinoma cells. Phytomedicine. 2018;**43**:37-45. DOI: 10.1016/j.phymed.2018.03.040

[19] Sajadi Hezaveh Z, Azarkeivan A, Janani L, Shidfar F. Effect of quercetin on oxidative stress and liver function in beta-thalassemia major patients receiving desferrioxamine: A double-blind randomized clinical trial. Journal of Research in Medical Sciences : The Official Journal of Isfahan University of Medical Sciences. 2019;**24**:91. DOI: 10.4103/jrms.JRMS_911_18

[20] Pasdar Y, Oubari F, Zarif MN, Abbasi M, Pourmahmoudi A, Hosseinikia M. Effects of quercetin supplementation on Hematological parameters in non-alcoholic fatty liver disease: A randomized, double-blind, placebo-controlled pilot study. Clinical Nutrition Research. 2020;**9**(1):11-19. DOI: 10.7762/cnr.2020.9.1.11

[21] Lu NT et al. A phase I dose escalation study demonstrates quercetin safety and explores potential for bioflavonoid antivirals in patients with chronic hepatitis C. Phytotherapy Research. 2016;**30**(1):160-168. DOI: 10.1002/ptr.5518

[22] Okamoto T. Safety of quercetin for clinical application (review). International Journal of Molecular Medicine. 2005;**16**(2):275-278

[23] Yi H et al. The therapeutic effects and mechanisms of quercetin on metabolic diseases: Pharmacological data and clinical evidence. Oxidative Medicine and Cellular Longevity. 2021;**2021**:6678662. DOI: 10.1155/2021/6678662

[24] Miles SL, McFarland M, Niles RM. Molecular and physiological actions of quercetin: Need for clinical trials to assess its benefits in human disease. Nutrition Reviews. 2014;**72**(11):720-734. DOI: 10.1111/nure.12152

[25] Chang TW. Binding of cells to matrixes of distinct antibodies coated on solid surface. Journal of Immunological Methods. 1983;**65**(1-2):217-223. DOI: 10.1016/0022-1759(83)90318-6

[26] Edgar R, Domrachev M, Lash AE. Gene expression omnibus: NCBI gene expression and hybridization array data repository. Nucleic Acids Research. 2002;**30**(1):207-210. DOI: 10.1093/nar/30.1.207

[27] Brazma A et al. Array express--a public repository for microarray gene expression data at the EBI. Nucleic Acids Research. 2003;**31**(1):68-71. DOI: 10.1093/nar/gkg091

[28] Brazma A et al. Minimum information about a microarray experiment (MIAME)-toward standards for microarray data. Nature Genetics. 2001;**29**(4):365-371. DOI: 10.1038/ng1201-365

[29] Mutch DM et al. Hepatic cytochrome P-450 reductase-null mice show reduced transcriptional response to quercetin and reveal physiological homeostasis between jejunum and liver. American Journal of Physiology. Gastrointestinal and Liver Physiology. 2006;**291**(1):G63-G72. DOI: 10.1152/ajpgi.00565.2005

[30] Rieswijk L, Brauers KJJ, Coonen MLJ, Jennen DGJ, van Breda SGJ, Kleinjans JCS. Exploiting microRNA and mRNA profiles generated in vitro from carcinogen-exposed primary mouse hepatocytes for predicting in vivo genotoxicity and carcinogenicity. Mutagenesis. 2016;**31**(5):603-615. DOI: 10.1093/mutage/gew027

[31] Amaya MJ, Nathanson MH. Calcium signaling in the liver. Comprehensive Physiology. 2013;**3**(1):515-539. DOI: 10.1002/cphy.c120013

[32] Dolphin AC. A short history of voltage-gated calcium channels. British Journal of Pharmacology. 2006;**147**(Suppl. 1):S56-S62. DOI: 10.1038/sj.bjp.0706442

[33] Dolphin AC. Voltage-gated calcium channels and their auxiliary subunits: Physiology and pathophysiology and pharmacology. The Journal of Physiology. 2016;**594**(19):5369-5390. DOI: 10.1113/JP272262

[34] Andrikopoulos P, Baba A, Matsuda T, Djamgoz MBA, Yaqoob MM, Eccles SA. Ca2+ influx through reverse mode Na+/Ca2+ exchange is critical for vascular endothelial growth factor-mediated extracellular signal-regulated kinase (ERK) 1/2 activation and angiogenic functions of human endothelial cells. The Journal of Biological Chemistry. 2011;**286**(44):37919-37931. DOI: 10.1074/jbc.M111.251777

[35] Wiel C et al. Endoplasmic reticulum calcium release through ITPR2 channels leads to mitochondrial calcium accumulation and senescence. Nature Communications. 2014;**5**:3792. DOI: 10.1038/ncomms4792

[36] Oliva-Vilarnau N, Hankeova S, Vorrink SU, Mkrtchian S, Andersson ER, Lauschke VM. Calcium Signaling in liver injury and regeneration. Frontiers in Medicine (Lausanne). 2018;**5**:192. DOI: 10.3389/fmed.2018.00192

[37] Dewidar B, Meyer C, Dooley S, Meindl-Beinker AN. TGF-β in hepatic stellate cell activation and liver Fibrogenesis-updated 2019. Cell. 2019;**8**(11):1419. DOI: 10.3390/cells8111419

[38] Zhang J, Jiang N, Ping J, Xu L. TGF β1 induced autophagy activates hepatic stellate cells via the ERK and JNK signaling pathways. International Journal of Molecular Medicine. 2021;**47**(1):256-266. DOI: 10.3892/ijmm.2020.4778

[39] Mezu-Ndubuisi OJ, Maheshwari A. The role of integrins in inflammation and angiogenesis. Pediatric Research. 2021;**89**(7):1619-1626. DOI: 10.1038/s41390-020-01177-9

[40] Aplin AE, Stewart SA, Assoian RK, Juliano RL. Integrin-mediated adhesion regulates ERK nuclear translocation and phosphorylation of Elk-1. The Journal of Cell Biology. 2001;**153**(2):273-282. DOI: 10.1083/jcb.153.2.273

Section 4

Antioxidant Properties

Chapter 5

Investigating the Antioxidant Properties of Quercetin

Kate Nyarko

Abstract

The antioxidant properties of quercetin stem from its ability to neutralize reactive oxygen species (ROS) and counteract oxidative stress, a key contributor to various chronic diseases. Numerous in vitro studies have demonstrated quercetin's effectiveness in scavenging free radicals and protecting cellular structures from oxidative damage. Beyond its direct antioxidant effects, quercetin also interacts with cellular signaling pathways, influencing gene expression and modulating enzymatic activities associated with oxidative stress. In vivo studies, both in animals and human trials, have provided insights into the bioavailability and physiological impact of quercetin, yet its significance remains underappreciated. This chapter will focus on the mechanisms by which quercetin enters circulation, its distribution in tissues, and the subsequent effects on markers of oxidative stress. Additionally, we will highlight findings from previous epidemiological studies linking quercetin-rich diets to reduced risk of chronic diseases, emphasizing the potential translational significance of these antioxidant properties in real-world health outcomes. In conclusion, this chapter will provide an overview of quercetin's antioxidant properties and its potential for therapeutic interventions associated with chronic diseases.

Keywords: antioxidant, quercetin, flavonoid, oxidative stress, chronic diseases

1. Introduction

Quercetin's chemical structure, characterized by the presence of multiple hydroxyl groups, provides it with potent free radical scavenging abilities, making it a key player in cellular defense against oxidative stress [1]. While there have been various reports indicating the potential therapeutic benefits of quercetin, its mechanism of action remains elusive and there is a dearth of clinical data to support its potential antioxidant properties in humans [2]. Quercetin, belonging to the flavonol subclass is widely distributed in nature and can be found in various food sources, including fruits vegetables and beverages. The availability of quercetin in commonly consumed dietary items has contributed to its accessibility and the growing interest in harnessing its potential health benefits. The structural arrangement of quercetin includes several hydroxyls (−OH) groups attached to different positions on the benzene rings, specifically at C3, C5, C7, C3′, C4′, and C5′ [3]. These hydroxyl groups are responsible for the antioxidant properties of quercetin, making it a potent scavenger of free radicals in the human body [4]. The conjugation of double bonds in the aromatic rings and the presence of hydroxyl groups further create a complex structure that plays a crucial role in quercetin's bioactivity and physiological effects. Additionally, the presence of a glycosyl group linked to the quercetin aglycone can affect its absorption, solubility, and in vivo effects [5]. Most studies focused on the biological effects of quercetin are based on in vitro research with quercetin aglycone, which is rarely found in human plasma [6, 7]. The findings from these studies have yielded inconclusive results regarding the exact mechanism of action of quercetin aglycone and its derived metabolites in humans [8]. Given the insufficient data in this research area, several authors have emphasized the importance of studying quercetin and their derived metabolites, considering that these metabolites may be the bioactive forms in the body, requiring further research to understand their mechanism of action and antioxidant activities [9–11]. In this study, our aim is to conduct a thorough investigation of the antioxidant capabilities of quercetin and its derivatives, and their impact on human health. Specifically, we will focus on their cellular signaling pathways, bioavailability, tissue distribution, and their association with reduced risk of chronic diseases.

2. Antioxidant role of quercetin in scavenging free radicals

Free radicals, including reactive oxygen species and reactive nitrogen species (RNS), are highly reactive molecules produced naturally during metabolic processes or by external factors like UV radiation, pollution, and certain chemicals. When present in excess, these free radicals can induce oxidative stress to cellular components such as proteins and lipids, leading to various chronic health conditions including neurodegenerative and cardiovascular diseases. Quercetin exerts its antioxidative effects through multiple mechanisms. Firstly, it acts as a direct scavenger of free radicals, neutralizing them and preventing them from causing harm to cellular structures as shown in **Figure 1**. Additionally, quercetin enhances the activity of endogenous antioxidant enzymes such as superoxide dismutase (SOD), catalase, and glutathione peroxidase, which further contribute to the cellular defense against oxidative stress.

Numerous in vitro and in vivo studies have highlighted the ability of quercetin to neutralize oxidative stress, making it a promising candidate for mitigating various chronic diseases associated with oxidative damage [12–15]. Various studies

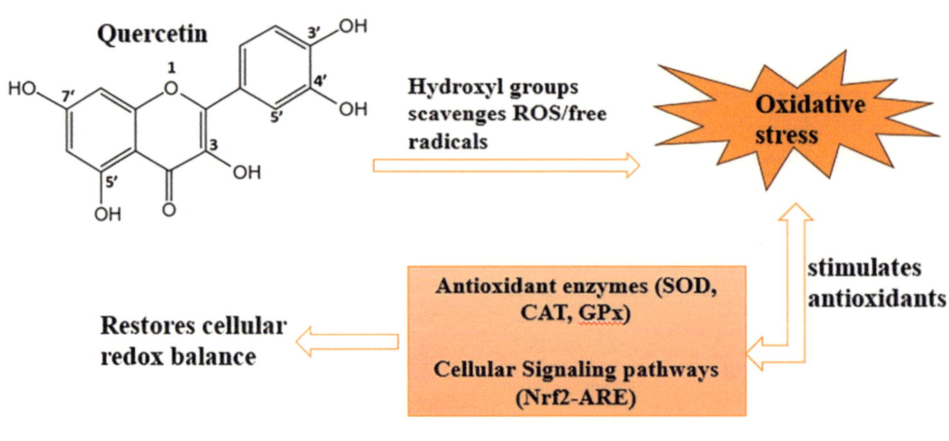

Figure 1.
The antioxidant activity of quercetin in scavenging free radicals.

have evaluated the antioxidant efficacy of quercetin and its derivatives using different assays, including 2,2-diphenyl-1-picrylhydrazyl (DPPH), Ferric Reducing Antioxidant Power (FRAP), Fe^{2+} chelation, and 2,2′-azino-bis (3-ethylbenzothiazoline-6-sulfonic acid) (ABTS) assays [16, 17]. The antioxidant activity of quercetin derivatives can be influenced by factors such as the acyl group chain length and the molar mass of the polymer. While the acylation of quercetin increases its lipophilicity, this can potentially influence its antioxidant efficacy. According to Oh et al. [18], the esterification of quercetin with various fatty acyl chlorides resulted in the formation of different derivatives, including Q-3′-O-monoester, Q-3-O-monoester, Q-7,3′-O-diester, Q-3′,4’-O- diester, Q-3,3′-O-diester, and Q-3,4′-O diester. The lipophilicity of these derivatives following the esterification process increased as expected. The results showed that quercetin had the highest activity in getting rid of harmful radicals compared to its derivatives. Interestingly, superior antioxidant activity was observed with the short-chain fatty acids derivatives in the ABTS assay, while those with C3:0, C4:0, C6:0, and C8:0 chains showed better performance in the DPPH radical scavenging assay.

In a previous study where the impact of nicotine and quercetin treatments on the viability and antioxidant defense system of HepG2 cell line cultures was investigated, quercetin showed an antioxidant protective effect by restoring the balance between free radical production and antioxidant defense systems [19]. The in vitro experimental results showed that quercetin increased the superoxide dismutase activity in both the control and nicotine treated group. The results were comparable to those from previous studies [20, 21]. Some studies have shown the ability of quercetin to combat oxidative stress by modulating glutathione (GSH) levels. For instance, the supplementation of quercetin in a Western diet increased the GSH/GSSG ratio in mice liver and mitigated obesity and metabolic syndrome [15]. The concentration of quercetin did not significantly influence body weight, fat storage, or blood composition in healthy mice following a standard diet. However, it did decrease oxidative stress levels, specifically in the liver. A higher quercetin concentration was observed to further decrease lipid peroxidation markers in various tissues. Additionally, quercetin's capacity to neutralize free radicals can be associated with a decrease in ischemia-reperfusion injury, achieved through alterations in the endothelial nitric oxide system [22]. This highlights the potential of quercetin to mitigate oxidative stress and protect against damage caused by ischemia-reperfusion processes. As an extension to this, recent research has emphasized quercetin's role in enhancing mitochondrial function [23, 24], further contributing to its overall protective effects on chronic and cardiovascular health diseases.

2.1 Absorption, distribution and metabolism of quercetin in tissues

The absorption of quercetin is a complex process influenced by several factors. Dietary intake is the primary source, and its absorption occurs in the small intestine. Quercetin, in its glycosidic form, undergoes hydrolysis by enzymes in the small intestine, releasing aglycones that are absorbed through passive diffusion [25]. Glucose transporters (GLUTs) and multidrug resistance-associated proteins (MRPs) are involved in the active transport of quercetin [26]. The release of aglycones from quercetin glycosides can be mediated by the lactase phlorizin hydrolase (LPH) in the small intestines [25]. Various factors impact the absorption of quercetin, including the food matrix, co-ingestion with other nutrients, and the gut microbiota. The presence of fats and certain nutrients enhances absorption, while high-fiber diets may reduce bioavailability. The bioavailability of quercetin can be improved when consumed alongside fatty acids in higher doses. Humans can absorb significant amounts from food or supplements, with a reported half-life of 11–28 hours. While daily quercetin intakes range from 3 to 40 mg in Western diets, consumers of fruits and vegetables may ingest about 250 mg per day. The recommended daily dose of quercetin supplements in diets should range from 500 to 1000 mg [27]. Additionally, the gut microbiota plays a crucial role in quercetin metabolism, influencing its absorption and bioactivity [28].

Once absorbed, quercetin is distributed to various tissues through the bloodstream. The distribution is influenced by its lipophilic nature, allowing it to penetrate cell membranes. On the other hand, quercetin exhibits a wide tissue distribution, with higher concentrations found in organs like the lungs, kidneys, and liver. The ability of quercetin to cross the blood-brain barrier has also raised interest in its potential neuroprotective effects [29, 30]. In the bloodstream and tissues, it undergoes extensive metabolism and biotransformation with phase II enzymes, such as glucuronosyltransferases and sulfotransferases, facilitating its excretion [31]. Special transporters also play a crucial role in facilitating the absorption and transport of quercetin and its metabolites. The lipid bilayers of cellular membranes, characterized by high polarity, limits the passage of quercetin conjugates where membrane-related transporters come to play to overcome this barrier. In the intestinal epithelial cells, numerous transporters are expressed to aid the passage of substrates from the gastrointestinal tract into the circulatory system [32]. Among these transporters, sodium-dependent glucose co-transporters (SGLTs) located on the apical membrane of intestinal epithelial cells play a significant role in the absorption of quercetin and its glycosides [33]. Specifically, quercetin 4′-β-glucoside and quercetin-3-glucoside are efficiently absorbed through the involvement of SGLT-1. There is evidence that the type of sugar linked to quercetin can influence the absorption ratio of quercetin glycosides in the small intestine [34]. Quercetin circulates in the bloodstream bound to plasma proteins, such as albumin [35]. The presence of binding proteins can influence its bioavailability and may impact its physiological effects. Understanding the binding characteristics of quercetin to plasma proteins is important for optimizing its delivery and enhancing its therapeutic potential. Further research is also needed to elucidate the specific mechanisms by which binding proteins interact with quercetin and how the interaction influences its bioavailability and physiological effects [36]. Furthermore, exploring strategies to enhance the bioavailability of quercetin, such as encapsulation in delivery systems like Quercetin LipoMicel, may improve its absorption and increase its circulating levels in the body [37].

2.2 Cellular signaling pathways of quercetin: modulation of enzymatic activities in oxidative stress

The pathway associated with the nuclear factor erythroid 2-related factor 2 (Nrf2) is vital for protecting cells from oxidative stress. Quercetin has been shown to activate the Nrf2 pathway, leading to the upregulation of antioxidant response element (ARE)-driven genes. This activation enhances the cellular antioxidant defense system, including the expression of enzymes like heme oxygenase-1 (HO-1) and superoxide dismutase. Quercetin and its derivative dihydroquercetin have demonstrated the ability to mitigate cellular damage induced by oxidative stress by activating the Nrf2-ARE pathway, with the involvement of stress-responsive Extracellular Signal-Regulated Kinase (ERK) and c-Jun N-terminal Kinase (JNK) signaling pathways, suggesting a potential neurohormetic role for quercetin as a phytochemical [14, 38, 39].

For instance, Jia et al. [40] examined the antioxidant effects of quercetin on diquat-induced oxidative stress in porcine enterocytes. Pretreatment with quercetin in intestinal porcine epithelial cell line 1 (IPEC-1) cells demonstrated protective effects by mitigating apoptosis through a caspase-3-dependent mechanism, reducing ROS production, maintaining mitochondrial function, and preserving tight junction proteins. The authors proposed that quercetin's protective role may be linked to the upregulation of Nrf2 protein levels and the modulation of intracellular glutathione content, highlighting its potential in regulating redox homeostasis. In a similar scenario, the activity of quercetin on p38-MAPK and its regulatory effects on the nuclear transcription factor erythroid-2p45-Nrf2 and the GSH antioxidant defense system in HepG2 cells was studied. The study showed a concentration-dependent modulation of p38 and Nrf2 over varying incubation periods, with 50 μM quercetin concentration [14]. In 2014, Saw and colleagues explored the cancer preventive properties of three flavonoids—quercetin, kaempferol, and pterostilbene—found in berries. The research focused on their ability to counteract free radicals using the DPPH and 2′,7′-Dichlorofluorescin diacetate (DCFH-DA) assays. The combination of all three flavonoids showed a synergistic effect in reducing intracellular ROS levels by stimulating the activation of ARE and the levels of mRNA and proteins associated with genes regulated by Nrf2 [38].

There is also evidence that quercetin can modulate enzymatic activities associated with oxidative stress [41–43]. One example is the inhibition of enzymes involved in ROS generation, such as NADPH oxidase [44–46]. By downregulating these enzymes, quercetin contributes to the reduction of intracellular ROS levels, thereby mitigating oxidative stress. In addition to that, quercetin interacts with key cellular signaling pathways, including mitogen-activated protein kinase (MAPK) and nuclear factor-kappa B (NF-κB) [47–49]. By inhibiting these pathways, quercetin plays a role in attenuating the expression of pro-inflammatory and pro-oxidative genes, contributing to its overall protective effects against oxidative stress-related damage. Furthermore, it modulates enzymatic activities involved in inflammatory pathways, including the inhibition of cyclooxygenase-2 (COX-2) and inducible nitric oxide synthase (iNOS) [50–52]. Quercetin was shown to markedly reduce the mRNA expression of inflammatory markers such as iNOS, COX-2, and C-reactive protein (CRP) in Chang Liver cells, demonstrating its anti-inflammatory effects [50].

2.3 Protective role of quercetin on chronic disease risk infection

Quercetin exerts its protective effects through diverse mechanisms. Its ability to neutralize ROS and free radicals arises from the presence of hydroxyl (−OH) groups and a catechol-type B-ring in its molecular structure. Additionally, quercetin activates the Nrf2 pathway, promoting the expression of antioxidant enzymes and cellular defense mechanisms. The interplay of these mechanism positions quercetin as a potent defender against oxidative stress-induced damage. Quercetin, known for its potent antioxidant properties, exhibits cardiovascular protection by reducing oxidative damage to endothelial cells, lowering blood pressure, and improving lipid profiles. Oxidative stress arises from an imbalance between ROS production and the body's antioxidant defense mechanisms [53, 54]. This imbalance is associated with the onset and progression of several chronic conditions such as myocardial infarction, chronic inflammation, aging, and neurodegenerative disorders [55, 56]. Numerous studies have demonstrated that quercetin exhibits neuroprotective effects and counteracts oxidative stress in vivo. For instance, its neuroprotective effect was shown in a study by Denny et al. [57] who found that the oral administration of quercetin and fish oil supplement enhanced neuroprotection in rats subjected to chronic exposure to the insecticide rotenone, which serves as an animal model for Parkinson's disease. Since the brain is particularly susceptible to oxidative stress, and a target for the damaging effects of ROS, the antioxidant effects of quercetin against oxidative stress pathways and its potential in preventing brain health diseases such as Alzheimer's disease (AD) through various pathways, including Nrf2, Paraoxonase-2, JNK, Protein kinase C, MAPK was discussed by Saikia et al. [58]. The study discussed that quercetin activates the Nrf2 pathway by increasing the production of protective proteins in AD brain cells, acting as a natural defense system against oxidative stress. Consequently, the neuroprotective potential of quercetin against oxidative stress responses was demonstrated by its interaction with paraoxonase-2-enzyme, JNK pathways and protein kinase C pathways which plays a role in cell signaling and function. Persistent activation of glutamate receptors, leading to excitotoxicity, contributes to various neurological disorders including Alzheimer's disease, hypoxic-ischemic brain injury, multiple sclerosis etc. [59]. However, quercetin extracts protect neuronal cells from excitotoxicity-induced damage through mechanisms involving reduced ROS production, preservation of mitochondrial membrane potential, and modulation of multiple biochemical markers associated with cell death and autophagy [60]. Quercetin has emerged as a significant component demonstrating acetylcholinesterase (AChE) inhibitory activity and exerts a positive influence on the expression of nicotinic receptors, thereby augmenting cognitive memory function in Alzheimer's patients [61–63].

Chronic inflammation is a prevalent contributing factor in various diseases, and quercetin's anti-inflammatory properties have been studied in the context of conditions such as rheumatoid arthritis and inflammatory bowel diseases [64]. Previous studies have suggested that quercetin-rich diets may be associated with a decreased risk of developing these inflammatory conditions, providing insights into potential dietary strategies for therapeutic interventions [65–67]. Its anti-inflammatory properties have been linked to potential cardiovascular benefits. While many studies suggest potential cardiovascular benefits of quercetin, it is important to note that research in this area is ongoing, and not all studies have produced consistent results. Factors such as dosage, bioavailability, and individual variability may influence quercetin's effects on cardiovascular health.

Excessive generation of reactive oxygen species resulting in oxidative stress has been recognized as a significant factor contributing to endothelial dysfunction-induced hypertension and various cardiovascular diseases [68]. Consequently, mitigating oxidative stress is considered a practical approach for comprehensive management of hypertension and other cardiovascular conditions. Several in vivo studies in animal models that explored the cardioprotective effects of quercetin and demonstrated its ability to lower blood pressure including those with hypertension, high-fat high-sucrose diet-induced conditions, nitric oxide deficiency, angiotensin infusion, and aortic constriction [69–72]. In a 4-week trial involving eighteen Dahl salt-sensitive rats, the antihypertensive effects of captopril (CAP) and quercetin (QUE) on the renin-angiotensin-aldosterone system were explored with a specific focus on renal effects. While the results from this study indicated that quercetin does not exhibit a significant difference in aldosterone levels, it effectively lowered blood pressure in the hypertensive rat model, suggesting a potential modulation of renal function as the underlying mechanism [72]. Hackl et al. [73] investigated the impact of quercetin on angiotensin-converting enzyme (ACE) activity by assessing cardiovascular responses to bradykinin and angiotensin I. Quercetin pretreatment, administered orally (88.7 umol/kg, 45 min) and intravenously (14.7 umol/kg, 5 min), significantly enhanced the hypotensive effect of bradykinin (10 nmol/kg, i.v.). This study indicated that quercetin possess an ACE inhibitory activity in vitro similar to captopril, demonstrating its potential antihypertensive properties.

In contrast to findings in animal studies, human research trials have not conclusively demonstrated consistent results regarding the antioxidant potential of quercetin, even at elevated dosages. Earlier studies by Egert et al. [8] and others [74–76] demonstrated no impact on plasma oxidized low-density lipoprotein in healthy individuals or those with pre-hypertension and stage 1 hypertension after quercetin supplementation. The discrepancy between the antioxidant effects observed in hypertensive animal models and the equivocal findings in humans may be attributed to variations in quercetin doses. However, Hertog and his group [77] provided promising results regarding the epidemiological and in vitro/in vivo antioxidant effects of flavonoids, supporting their cardioprotective function. Their study provided evidence supporting the strong cardioprotective effects of various flavonoids, including quercetin. The study showed that men who consumed over 29 mg of flavonols per day experienced a substantial 68% reduction in the risk of coronary death compared to those consuming less than 10 mg per day. Although the study did not specifically explore the link between quercetin intake and blood pressure, the authors noted an inverse relationship between high-quercetin foods and blood pressure, and flavonol intake, including quercetin. While these studies conducted in animal models have indicated a decrease in blood pressure and enhanced antioxidant status with quercetin administration, the evidence from human trials has not been particularly compelling. Human studies examining the effects of quercetin on blood pressure and antioxidant status have not yielded convincing results, indicating the need for additional investigation and exploration of potential factors that influence the outcomes in human subjects. The translation of promising animal findings to human contexts requires careful consideration of various variables that may contribute to the differences observed in the effects of quercetin on blood pressure and antioxidant status.

Quercetin's ability to scavenge free radicals and modulate inflammatory pathways suggests that it may contribute to reducing oxidative stress and inflammation associated with diabetes, thereby potentially alleviating the risk of diabetic complications. Quercetin's potential protective effects on pancreatic beta-cells, responsible for insulin production, have been extensively investigated [78]. Various studies have shown that quercetin may protect beta-cells from oxidative stress and apoptosis, thereby preserving their function. Furthermore, quercetin may stimulate insulin secretion, contributing to better glycemic control. Additionally, it may inhibit key enzymes involved in carbohydrate digestion, leading to a more gradual release of glucose into the bloodstream. Various in vitro studies have focused on the antidiabetic effects of quercetin in cellular and animal models. For example, the impact of quercetin on glucose or glibenclamide-induced insulin secretion and its ability to protect against hydrogen peroxide-induced beta-cell dysfunctions, using the INS-1 beta-cell line was examined. The researchers observed that quercetin enhances insulin secretion, protects beta-cells from oxidative damage, and implicates the extracellular signal-regulated kinase (ERK)1/2 pathway [79]. Additionally, quercetin exhibited concentration-dependent inhibition of alpha-amylase and alpha-glucosidase activities. These findings suggest that quercetin's ability to regulate blood glucose levels may be attributed to its disruption of enzymatic processes and protection of pancreatic tissues from oxidative damage, as evidenced by the inhibition of carbohydrate-metabolizing enzymes and prevention of pancreatic lipid peroxidation [80]. Moreover, potential interactions with medications and variations in individual responses should be considered when assessing the overall safety profile of quercetin for individuals with diabetes.

3. Conclusions

In conclusion, quercetin, with its powerful antioxidant properties, is crucial for defending cells against oxidative stress. While much research has focused on its potential therapeutic benefits, clinical evidence supporting its antioxidant effects in humans is limited. Understanding the mechanisms of action of quercetin and its derivatives is crucial, especially considering that these metabolites may be the active forms in the body. Despite challenges in absorption and bioavailability, quercetin's ability to mitigate oxidative damage holds promise for combating various chronic diseases. Additionally, its activation of the Nrf2 pathway and interaction with signaling pathways further underscores its potential as a neuroprotective agent. Further research is needed to fully elucidate the mechanisms underlying quercetin's antioxidant activities including clinical trials to fully elucidate the therapeutic potential and optimal dosage of quercetin for specific health conditions.

Author details

Kate Nyarko
University of Missouri-Columbia, Columbia, USA

*Address all correspondence to: knzhb@umsystem.edu

References

[1] Srimathi Priyanga K, Vijayalakshmi K. Investigation of antioxidant potential of quercetin and hesperidin: An in vitro approach. Asian Journal of Pharmaceutical and Clinical Research. 2017;**10**:83-86. DOI: 10.22159/AJPCR.2017.V10I11.20260

[2] Hagde P, Pingle P, Mourya A, Katta CB, Srivastava S, Sharma R, et al. Therapeutic potential of quercetin in diabetic foot ulcer: Mechanistic insight, challenges, nanotechnology driven strategies and future prospects. Journal of Drug Delivery Science and Technology. 2022;**74**:103575. DOI: 10.1016/J.JDDST.2022.103575

[3] Li Y, Yao J, Han C, Yang J, Chaudhry MT, Wang S, et al. Quercetin, inflammation and immunity. Nutrients. 2016;**8**:167. DOI: 10.3390/NU8030167

[4] Ozgen S, Kilinc OK, Selamoğlu Z. Antioxidant activity of quercetin: A mechanistic review. Turkish Journal of Agriculture-Food Science and Technology. 2016;**4**:1134-1138

[5] Wang W, Sun C, Mao L, Ma P, Liu F, Yang J, et al. The biological activities, chemical stability, metabolism and delivery systems of quercetin: A review. Trends in Food Science and Technology. 2016;**56**:21-38. DOI: 10.1016/J.TIFS.2016.07.004

[6] Kroon PA, Clifford MN, Crozier A, Day AJ, Donovan JL, Manach C, et al. How should we assess the effects of exposure to dietary polyphenols in vitro? The American Journal of Clinical Nutrition. 2004;**80**:15-21. DOI: 10.1093/AJCN/80.1.15

[7] Mullen W, Edwards CA, Crozier A. Absorption, excretion and metabolite profiling of methyl-, glucuronyl-, glucosyl- and sulpho-conjugates of quercetin in human plasma and urine after ingestion of onions. British Journal of Nutrition. 2006;**96**:107. DOI: 10.1079/BJN20061809

[8] Egert S, Wolffram S, Bosy-Westphal A, Boesch-Saadatmandi C, Wagner AE, Frank J, et al. Daily quercetin supplementation dose-dependently increases plasma quercetin concentrations in healthy humans. The Journal of Nutrition. 2008;**138**:1615-1621. DOI: 10.1093/JN/138.9.1615

[9] Wiczkowski W, Szawara-Nowak D, Topolska J, Olejarz K, Zieliński H, Piskuła MK. Metabolites of dietary quercetin: Profile, isolation, identification, and antioxidant capacity. Journal of Functional Foods. 2014;**11**:121-129. DOI: 10.1016/J.JFF.2014.09.013

[10] Santos MR, Rodríguez-Gómez MJ, Justino GC, Charro N, Florencio MH, Mira L. Influence of the metabolic profile on the in vivo antioxidant activity of quercetin under a low dosage oral regimen in rats. British Journal of Pharmacology. 2008;**153**:1750. DOI: 10.1038/BJP.2008.46

[11] Cho JM, Chang SY, Kim DB, Needs PW, Jo YH, Kim MJ. Effects of physiological quercetin metabolites on interleukin-1β-induced inducible NOS expression. The Journal of Nutritional Biochemistry. 2012;**23**:1394-1402. DOI: 10.1016/J.JNUTBIO.2011.08.007

[12] Ademosun AO, Oboh G, Bello F, Ayeni PO. Antioxidative properties and effect of quercetin and its glycosylated form (Rutin) on acetylcholinesterase and butyrylcholinesterase activities. Evidence-based

Complementary and Alternative Medicine. 2016;**21**:NP11-NP17. DOI: 10.1177/2156587215610032/ASSET/IMAGES/LARGE/10.1177_2156587215610032-FIG7.JPEG

[13] Braun KF, Ehnert S, Freude T, Egaña JT, Schenck TL, Buchholz A, et al. Quercetin protects primary human osteoblasts exposed to cigarette smoke through activation of the antioxidative enzymes HO-1 and SOD-1. Research article. The Scientific World Journal. 2011;**11**:2348-2357. DOI: 10.1100/2011/471426

[14] Granado-Serrano AB, Martín MA, Bravo L, Goya L, Ramos S. Quercetin modulates Nrf2 and glutathione-related defenses in HepG2 cells: Involvement of P38. Chemico-Biological Interactions. 2012;**195**:154-164. DOI: 10.1016/J.CBI.2011.12.005

[15] Kobori M, Takahashi Y, Akimoto Y, Sakurai M, Matsunaga I, Nishimuro H, et al. Chronic high intake of quercetin reduces oxidative stress and induces expression of the antioxidant enzymes in the liver and visceral adipose tissues in mice. Journal of Functional Foods. 2015;**15**:551-560. DOI: 10.1016/J.JFF.2015.04.006

[16] Murugesan N, Damodaran C, Krishnamoorthy S, Raja M. In-vitro evaluation of synergism in antioxidant efficiency of quercetin and resveratrol. Chemical Biology Letters. 2023;**10**:534-534

[17] Zizkova P, Stefek M, Rackova L, Prnova M, Horakova L. Novel quercetin derivatives: From redox properties to promising treatment of oxidative stress related diseases. Chemico-Biological Interactions. 2017;**265**:36-46. DOI: 10.1016/J.CBI.2017.01.019

[18] Oh WY, Ambigaipalan P, Shahidi F. Preparation of quercetin esters and their antioxidant activity. Journal of Agricultural and Food Chemistry. 2019;**67**:10653-10659. DOI: 10.1021/ACS.JAFC.9B04154

[19] Yarahmadi A, Zal F, Bolouki A. Protective effects of quercetin on nicotine induced oxidative stress in 'HepG2 cells'. Toxicology Mechanisms and Methods. 2017;**27**:609-614. DOI: 10.1080/15376516.2017.1344338

[20] Lee YJ, Beak SY, Choi I, Sung JS. Quercetin and its metabolites protect hepatocytes against ethanol-induced oxidative stress by activation of Nrf2 and AP-1. Food Science and Biotechnology. 2018;**27**:809. DOI: 10.1007/S10068-017-0287-8

[21] Muthukumaran S, Sudheer AR, Menon VP, Nalini N. Protective effect of quercetin on nicotine-induced prooxidant and antioxidant imbalance and DNA damage in Wistar rats. Toxicology. 2008;**243**:207-215. DOI: 10.1016/J.TOX.2007.10.006

[22] Shoskes D, Lapierre C, Cruz-Corerra M, Muruve N, Rosario R, Fromkin B, et al. Beneficial effects of the bioflavonoids curcumin and quercetin on early function in cadaveric renal transplantation: A randomized placebo controlled trial. Transplantation. 2005;**80**:1556-1559. DOI: 10.1097/01.TP.0000183290.64309.21

[23] Qiu L, Luo Y, Chen X. Quercetin attenuates mitochondrial dysfunction and biogenesis via upregulated AMPK/SIRT1 signaling pathway in OA rats. Biomedicine & Pharmacotherapy. 2018;**103**:1585-1591. DOI: 10.1016/J.BIOPHA.2018.05.003

[24] Rayamajhi N, Kim SK, Go H, Joe Y, Callaway Z, Kang JG, et al. Quercetin induces mitochondrial biogenesis through activation of HO-1 in

HepG2 cells. Oxidative Medicine and Cellular Longevity. 2013;**2013**:2-10. DOI: 10.1155/2013/154279

[25] Chen L, Cao H, Huang Q, Xiao J, Teng H. Absorption, metabolism and bioavailability of flavonoids: A review. Critical Reviews in Food Science and Nutrition. 2022;**62**:7730-7742. DOI: 10.1080/10408398.2021.1917508

[26] Wenzel U. Flavonoids as drugs at the small intestinal level. Current Opinion in Pharmacology. 2013;**13**:864-868. DOI: 10.1016/J.COPH.2013.08.015

[27] Andres S, Pevny S, Ziegenhagen R, Bakhiya N, Schäfer B, Hirsch-Ernst KI, et al. Safety aspects of the use of quercetin as a dietary supplement. Molecular Nutrition & Food Research. 2018;**62**:2-17. DOI: 10.1002/MNFR.201700447

[28] Murota K, Nakamura Y, Uehara M. Flavonoid metabolism: The interaction of metabolites and gut microbiota. Bioscience, Biotechnology, and Biochemistry. 2018;**82**:600-610. DOI: 10.1080/09168451.2018.1444467

[29] Grabowska D, Galanty K, Sobolewska A, Podolak D, Saldanha L, Fernanda De Campos Fraga-Silva T, et al. A flavonoid on the brain: Quercetin as a potential therapeutic agent in central nervous system disorders. Life. 2022;**12**:591. DOI: 10.3390/LIFE12040591.

[30] Grewal AK, Singh TG, Sharma D, Sharma V, Singh M, Rahman MH, et al. Mechanistic insights and perspectives involved in neuroprotective action of quercetin. Biomedicine & Pharmacotherapy. 2021;**140**:111729. DOI: 10.1016/J.BIOPHA.2021.111729

[31] Jiang W, Hu M. Mutual interactions between flavonoids and enzymatic and transporter elements responsible for flavonoid disposition via phase II metabolic pathways. RSC Advances. 2012;**2**:7948-7963. DOI: 10.1039/C2RA01369J

[32] Han F, Yang P, Wang H, Fernandes I, Mateus N, Liu Y. Digestion and absorption of red grape and wine anthocyanins through the gastrointestinal tract. Trends in Food Science and Technology. 2019;**83**:211-224. DOI: 10.1016/J.TIFS.2018.11.025

[33] Röder PV, Geillinger KE, Zietek TS, Thorens B, Koepsell H, Daniel H. The role of SGLT1 and GLUT2 in intestinal glucose transport and sensing. PLoS One. 2014;**9**:2-10. DOI: 10.1371/JOURNAL.PONE.0089977

[34] Arts ICW, Sesink ALA, Faassen-Peters M, Hollman PCH. The type of sugar moiety is a major determinant of the small intestinal uptake and subsequent biliary excretion of dietary quercetin glycosides. The British Journal of Nutrition. 2004;**91**:841-847. DOI: 10.1079/BJN20041123

[35] Aghababaei F, Hadidi M. Recent advances in potential health benefits of quercetin. Pharmaceuticals. 2023;**16**:2-31. DOI: 10.3390/PH16071020

[36] Papakyriakopoulou P, Velidakis N, Khattab E, Valsami G, Korakianitis I, Kadoglou NPE. Potential pharmaceutical applications of quercetin in cardiovascular diseases. Pharmaceuticals. 2022;**15**:1019-1019. DOI: 10.3390/PH15081019

[37] Solnier J, Chang C, Roh K, Du M, Kuo YC, Hardy M, et al. Quercetin LipoMicel—A novel delivery system to enhance bioavailability of quercetin. Journal of Natural Health Product Research. 2021;**3**:1-8. DOI: 10.33211/JNHPR.17

[38] Saw CLL, Guo Y, Yang AY, Paredes-Gonzalez X, Ramirez C, Pung D, et al. The berry constituents quercetin, kaempferol, and pterostilbene synergistically attenuate reactive oxygen species: Involvement of the Nrf2-ARE signaling pathway. Food and Chemical Toxicology. 2014;**72**:303-311. DOI: 10.1016/J.FCT.2014.07.038

[39] Liang L, Gao C, Luo M, Wang W, Zhao C, Zu Y, et al. Dihydroquercetin (DHQ) induced HO-1 and NQO1 expression against oxidative stress through the Nrf2-dependent antioxidant pathway. Journal of Agricultural and Food Chemistry. 2013;**61**:2755-2761. DOI: 10.1021/JF304768P/ASSET/IMAGES/LARGE/JF-2012-04768P_0006.JPEG

[40] Jia H, Zhang Y, Si X, Jin Y, Jiang D, Dai Z, et al. Quercetin alleviates oxidative damage by activating nuclear factor erythroid 2-related factor 2 signaling in porcine enterocytes. Nutrients. 2021;**13**:1-15. DOI: 10.3390/NU13020375

[41] Costa LG, Garrick JM, Roquè PJ, Pellacani C. Mechanisms of neuroprotection by quercetin: Counteracting oxidative stress and more. Oxidative Medicine and Cellular Longevity. 2016;**2016**:1-10. DOI: 10.1155/2016/2986796

[42] Ramyaa P, Krishnaswamy R, Padma VV. Quercetin modulates OTA-induced oxidative stress and redox signalling in HepG2 cells — Up regulation of Nrf2 expression and down regulation of NF-KB and COX-2. Biochimica et Biophysica Acta (BBA) - General Subjects. 2014;**1840**:681-692. DOI: 10.1016/J.BBAGEN.2013.10.024

[43] Alía M, Ramos S, Mateos R, Granado-Serrano AB, Bravo L, Goya L. Quercetin protects human hepatoma HepG2 against oxidative stress induced by tert-butyl hydroperoxide. Toxicology and Applied Pharmacology. 2006;**212**:110-118. DOI: 10.1016/J.TAAP.2005.07.014

[44] Baba RA, Mir HA, Mokhdomi TA, Bhat HF, Ahmad A, Khanday FA. Quercetin suppresses ROS production and migration by specifically targeting Rac1 activation in gliomas. Frontiers in Pharmacology. 2024;**15**:1318797. DOI: 10.3389/FPHAR.2024.1318797

[45] Sul OJ, Ra SW. Quercetin prevents LPS-induced oxidative stress and inflammation by modulating NOX2/ROS/NF-KB in lung epithelial cells. Molecules. 2021;**22**:1-11. DOI: 10.3390/MOLECULES26226949

[46] Luo M, Tian R, Yang Z, Peng YY, Lu N. Quercetin suppressed NADPH oxidase-derived oxidative stress via heme oxygenase-1 induction in macrophages. Archives of Biochemistry and Biophysics. 2019;**671**:69-76. DOI: 10.1016/J.ABB.2019.06.007

[47] Devi KP, Kiruthiga P, Pandian SK. Emerging role of flavonoids in inhibition of NF-B-mediated signaling pathway: A review. International Journal of Biomedical and Pharmaceutical Sciences. 2009;**3**:31-45

[48] Endale M, Park SC, Kim S, Kim SH, Yang Y, Cho JY, et al. Quercetin disrupts tyrosine-phosphorylated phosphatidylinositol 3-kinase and myeloid differentiation factor-88 association, and inhibits MAPK/AP-1 and IKK/NF-KB-induced inflammatory mediators production in RAW 264.7 cells. Immunobiology. 2013;**218**:1452-1467. DOI: 10.1016/J.IMBIO.2013.04.019

[49] Ruiz PA, Braune A, Hölzlwimmer G, Quintanilla-Fend L, Haller D. Quercetin inhibits TNF-induced NF-KappaB transcription factor recruitment to

proinflammatory gene promoters in murine intestinal epithelial cells. The Journal of Nutrition. 2007;**137**:1208-1215. DOI: 10.1093/JN/137.5.1208

[50] García-Mediavilla V, Crespo I, Collado PS, Esteller A, Sánchez-Campos S, Tuñón MJ, et al. The anti-inflammatory flavones quercetin and kaempferol cause inhibition of inducible nitric oxide synthase, cyclooxygenase-2 and reactive C-protein, and down-regulation of the nuclear factor KappaB pathway in Chang liver cells. European Journal of Pharmacology. 2007;**557**:221-229. DOI: 10.1016/J.EJPHAR.2006.11.014

[51] Chen YC, Shen SC, Lee WR, Hou WC, Yang LL, Lee TJF. Inhibition of nitric oxide synthase inhibitors and lipopolysaccharide induced inducible NOS and cyclooxygenase-2 gene expressions by Rutin, quercetin, and quercetin pentaacetate in RAW 264.7 macrophages. Journal of Cellular Biochemistry. 2001;**82**:537-548. DOI: 10.1002/JCB.1184

[52] Raso GM, Meli R, Di Carlo G, Pacilio M, Di Carlo R. Inhibition of inducible nitric oxide synthase and cyclooxygenase-2 expression by flavonoids in macrophage J774A.1. Life Sciences. 2001;**68**:921-931. DOI: 10.1016/S0024-3205(00)00999-1

[53] Dong YS, Wang JL, Feng DY, Qin HZ, Wen H, Yin ZM, et al. Protective effect of quercetin against oxidative stress and brain edema in an experimental rat model of subarachnoid hemorrhage. International Journal of Medical Sciences. 2014;**11**:282. DOI: 10.7150/IJMS.7634

[54] Betteridge DJ. What is oxidative stress? Metabolism. 2000;**49**:3-8. DOI: 10.1016/S0026-0495(00)80077-3

[55] Uttara B, Singh AV, Zamboni P, Mahajan RT. Oxidative stress and neurodegenerative diseases: A review of upstream and downstream antioxidant therapeutic options. Current Neuropharmacology. 2009;7:65. DOI: 10.2174/157015909787602823

[56] Fridovich I. Fundamental aspects of reactive oxygen species, or what's the matter with oxygen? Annals of the New York Academy of Sciences. 1999;**893**:13-18. DOI: 10.1111/J.1749-6632. 1999.TB07814.X

[57] Denny Joseph KM, Muralidhara. Combined oral supplementation of fish oil and quercetin enhances neuroprotection in a chronic rotenone rat model: Relevance to Parkinson's disease. Neurochemical Research. 2015;**40**:894-905. DOI: 10.1007/S11064-015-1542-0

[58] Saikia S. Ascertaining the antioxidant properties of quercetin against oxidative stress in combating Alzheimer's disease: A review. Uttar Pradesh Journal of Zoology. 2022;**43**:680-694. DOI: 10.56557/UPJOZ/2022/V43I243385

[59] Akyuz E, Paudel YN, Polat AK, Dundar HE, Angelopoulou E. Enlightening the neuroprotective effect of quercetin in epilepsy: From mechanism to therapeutic opportunities. Epilepsy & Behavior. 2021;**115**:1525-5050. DOI: 10.1016/J.YEBEH.2020.107701

[60] Silva B, Oliveira PJ, Dias A, Malva JO. Quercetin, kaempferol and Biapigenin from hypericum Perforatum are neuroprotective against excitotoxic insults. Neurotoxicity Research. 2008;**13**:265-279. DOI: 10.1007/BF03033510

[61] Parsons CG, Danysz W, Dekundy A, Pulte I. Memantine and cholinesterase inhibitors: Complementary mechanisms in the treatment of Alzheimer's disease. Neurotoxicity Research. 2013;**24**:358. DOI: 10.1007/S12640-013-9398-Z

[62] Islam MR, Zaman A, Jahan I, Chakravorty R, Chakraborty S. In silico QSAR analysis of quercetin reveals its potential as therapeutic drug for Alzheimer's disease. Journal of Young Pharmacists. 2013;**5**:173. DOI: 10.1016/J.JYP.2013.11.005

[63] Jung M, Park M. Acetylcholinesterase inhibition by flavonoids from Agrimonia Pilosa. Molecules: A Journal of Synthetic Chemistry and Natural Product Chemistry. 2007;**12**:2130. DOI: 10.3390/12092130

[64] El-Said KS, Atta A, Mobasher MA, Germoush MO, Mohamed TM, Salem MM. Quercetin mitigates rheumatoid arthritis by inhibiting adenosine deaminase in rats. Molecular Medicine. 2022;**28**:1-13. DOI: 10.1186/S10020-022-00432-5/FIGURES/11

[65] Habtemariam S, Belai A. Natural therapies of the inflammatory bowel disease: The case of Rutin and its aglycone, quercetin. Mini-reviews in Medicinal Chemistry. 2017;**18**(3):234-243. DOI: 10.2174/1389557517666170120152417

[66] Comalada M, Camuesco D, Sierra S, Ballester I, Xaus J, Gálvez J, et al. In vivo quercitrin anti-inflammatory effect involves release of quercetin, which inhibits inflammation through down-regulation of the NF-KB pathway. European Journal of Immunology. 2005;**35**:584-592. DOI: 10.1002/EJI.200425778

[67] Kleemann R, Verschuren L, Morrison M, Zadelaar S, van Erk MJ, Wielinga PY, et al. Anti-inflammatory, anti-proliferative and anti-atherosclerotic effects of quercetin in human in vitro and in vivo models. Atherosclerosis. 2011;**218**:44-52. DOI: 10.1016/J.ATHEROSCLEROSIS.2011.04.023

[68] Schiffrin EL. Antioxidants in hypertension and cardiovascular disease. Molecular Interventions. 2010;**10**:354-362. DOI: 10.1124/MI.10.6.4

[69] Yamamoto Y, Oue E. Antihypertensive effect of quercetin in rats fed with a high-fat high-sucrose diet. Bioscience, Biotechnology, and Biochemistry. 2006;**70**:933-939. DOI: 10.1271/BBB.70.933

[70] Duarte J, Jiménez R, O'Valle F, Galisteo M, Pérez-Palencia R, Vargas F, et al. Protective effects of the flavonoid quercetin in chronic nitric oxide deficient rats. Journal of Hypertension. 2002;**20**:1843-1854. DOI: 10.1097/00004872-200209000-00031

[71] Jalili T, Carlstrom J, Kim S, Freeman D, Jin H, Wu TC, et al. Quercetin-supplemented diets lower blood pressure and attenuate cardiac hypertrophy in rats with aortic constriction. Journal of Cardiovascular Pharmacology. 2006;**47**:531-541. DOI: 10.1097/01.FJC.0000211746.78454.50

[72] Mackraj I, Govender T, Ramesar S. The antihypertensive effects of quercetin in a salt-sensitive model of hypertension. Journal of Cardiovascular Pharmacology. 2008;**51**:239-245. DOI: 10.1097/FJC.0B013E318162011F

[73] Häckl LPN, Cuttle G, Sanches Dovichi S, Lima-Landman MT, Nicolau M. Inhibition of angiotensin-converting enzyme by quercetin alters the vascular response to bradykinin and angiotensin I. Pharmacology. 2002;**65**:182-186. DOI: 10.1159/000064341

[74] Larson AJ, Symons JD, Jalili T. Quercetin: A treatment for hypertension?—A review of efficacy and mechanisms. Pharmaceuticals. 2010;**3**:237. DOI: 10.3390/PH3010237

[75] Conquer JA, Maiani G, Azzini E, Raguzzini A, Holub BJ. Supplementation with quercetin markedly increases plasma quercetin concentration without effect on selected risk factors for heart disease in healthy subjects. The Journal of Nutrition. 1998;**128**:593-597. DOI: 10.1093/JN/128.3.593

[76] Edwards RL, Lyon T, Litwin SE, Rabovsky A, Symons JD, Jalili T. Quercetin reduces blood pressure in hypertensive subjects. The Journal of Nutrition. 2007;**137**:2405-2411. DOI: 10.1093/JN/137.11.2405

[77] Hertog MGL, Feskens EJM, Kromhout D, Hertog MGL, Hollman PCH, Hertog MGL, et al. Dietary antioxidant flavonoids and risk of coronary heart disease: The Zutphen elderly study. The Lancet. 1993;**342**:1007-1011. DOI: 10.1016/0140-6736(93)92876-U

[78] Wang JY, Nie YX, Dong BZ, Cai ZC, Zeng XK, Du L, et al. Quercetin protects islet β-cells from oxidation-induced apoptosis via Sirt3 in T2DM. Iranian Journal of Basic Medical Sciences. 2021;**24**:629. DOI: 10.22038/IJBMS.2021.52005.11792

[79] Youl E, Bardy G, Magous R, Cros G, Sejalon F, Virsolvy A, et al. Quercetin potentiates insulin secretion and protects INS-1 pancreatic b-cells against oxidative damage via the ERK1/2 pathway. British Journal of Pharmacology. 2010;**161**(4):799-814. DOI: 10.1111/j.1476-5381.2010.00910.x

[80] Oboh G, Ademosun AO, Ayeni PO, Omojokun OS, Bello F. Comparative effect of quercetin and Rutin on α-amylase, α-glucosidase, and some pro-oxidant-induced lipid peroxidation in rat pancreas. Comparative Clinical Pathology. 2015;**24**:1103-1110. DOI: 10.1007/S00580-014-2040-5

Section 5

Therapeutic Applications

Chapter 6

Antidiabetic Potential of Quercetin

Priya Mijgar and Uday Deokate

Abstract

Natural flavonoid quercetin is widely distributed in fruits and vegetables. Quercetin may be therapeutically useful in the treatment and prevention of a variety of diseases, including cancer, diabetes, and cardiovascular disease, according to mounting research. The aim of the chapter is to provide information on the antidiabetic properties of quercetin. According to studies, quercetin regulates whole-body glucose homeostasis through interacting with a variety of molecular targets in the liver, pancreas, skeletal muscle, adipose tissue, and small intestine. The pleiotropic mechanisms of action of quercetin include the reduction of intestinal glucose absorption, insulin secretory, and insulin-sensitizing actions, as well as enhanced glucose utilization in peripheral tissues.

Keywords: quercetin, flavonoids, diabetes, gluconeogenesis, glucose absorption, glucose uptake, insulin resistance, insulin secretion, AMPK

1. Introduction

The prevalence of diabetes mellitus (DM), which has been steadily rising over time, is a serious issue that affects the entire world. Worldwide, especially in low- to middle-income nations, the prevalence of DM has been rising. In the previous 30 years, from 108 million in 1980 to 422 million in 2014, the number of persons with DM has nearly quadrupled, according to a global report from the World Health Organization (WHO). Over 1.5 million deaths in 2012 were specifically related to DM, and 2.2 million deaths in 2012 were caused by hyperglycemia. DM is expected to rank as the sixth leading cause of mortality worldwide by the year 2030 [1]. DM has also been causing severe organ failure over the years rapidly becoming one of the noncommunicable diseases, causing a rapid increase in mortality rates [2]. The two main types of diabetes are insulin-dependent diabetes mellitus (IDDM, type 1) and noninsulin-dependent diabetes mellitus (NIDDM, type 2), as shown in **Figure 1**. Diabetes mellitus type 2 are caused by a complex metabolic disorder in the endocrine system, which is characterized by hyperglycemia and obstruction in the metabolism of protein, carbohydrates, and lipid as a result of either impairment or destruction of insulin action and insulin-secreting pancreatic cells in target tissues. The insulin-secreting cells in type 11 diabetes, an autoimmune condition, were targeted and destroyed by regional inflammatory mediators [3, 4]. The most encountered form of diabetes is NIDDM as it accounts for more than 80% of the total case of DM hyperglycemia brought on by impaired insulin production or insulin resistance characterizes diabetes as an endocrine illness. Diabetes is primarily divided into

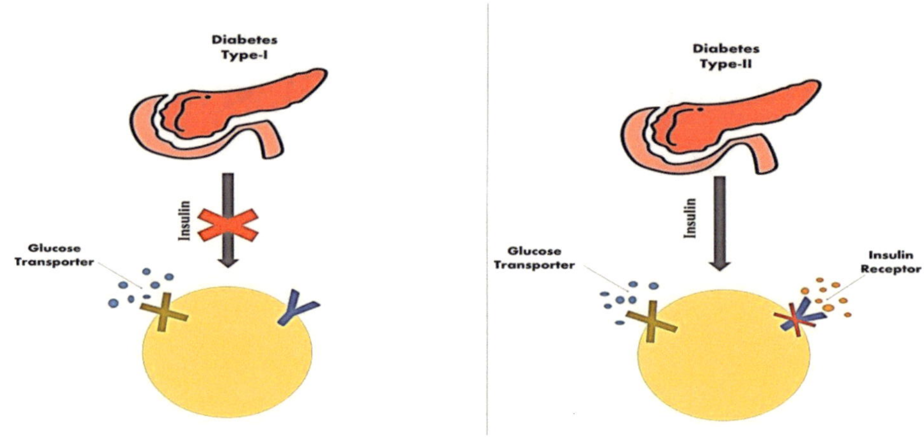

Figure 1.
Types of diabetes mellitus.

type I and type II, as depicted in **Figure 1**. Type I diabetes, commonly known as juvenile diabetes due to its predominance in youngsters, is brought on by flaws in the pancreatic insulin secretion process. As depicted in **Figure 1**, type II diabetes (T2DM) is characterized by insulin resistance, in which insulin binds to its receptor in the cell membrane without activating the signaling cascade necessary for glucose metabolism. It leads to pancreatic β-cell failure, insulin resistance, and progressive hyperglycemia [5]. In addition, chronic hyperglycemia caused by DM has become a serious problem as it is associated with serious long-term health complications such as dysfunction and failure of many organs, including nerves, heart, blood vessels, kidneys, and eyes [6]. One of the many therapeutic approaches that decrease postprandial hyperglycemia is to inhibit the carbohydrate-hydrolyzing enzyme, glucosidase, found in the intestine that plays a huge role in facilitating the uptake of glucose by breaking down carbohydrates before monosaccharide absorption. Alpha-glucosidase inhibitors also play an important role in reducing the insulin peaks and postprandial glycemia by delaying the absorption of ingested carbohydrates [7]. In recent decades, researchers have been focused on developing therapeutic agents or strategies to address the pathological aspects of the disease. Diabetes is a progressive disease that requires multiple drugs, frequently in addition to insulin. Unfortunately, most of the currently used drugs have side effects and some bioflavonoids mechanism action, as shown in **Tables 1** and **2**, respectively [8, 9].

The acceleration of population aging, as well as behavioral and socioeconomic shifts, have made metabolic illnesses a significant public health issue Therefore, the search for safe and efficient treatment medications is urgent. Physiotherapy has been valued in diverse traditional cultures, and it is believed that natural products are more economical and safer than chemical products. In the prevention or/and treatment of diseases, natural substances derived from fruits, vegetables, or medicinal plants are frequently used. Flavonoids are a superfamily of phytochemicals that have been taken from natural plants and are thought to offer therapeutic promise for metabolic disorders [10]. Quercetin is a crucial flavonoid component that is present in a variety of edible and therapeutic plants. Due to its excellent potential for the treatment of metabolic illnesses, quercetin has attracted a lot of interest recently [11]. Numerous

Therapeutic agent	Molecular action	Adverse effects
Metformin (Biguanides)	AMPK pathway	Felling seek, gastrointestinal disturbances, nausea, vomiting.
Insulin	Insulin receptors in muscle, Liver, adipose tissue	Hypoglycemia, blurred vision, fast heartbeat
Thiazolidinediones	PPAR-γ, GLUT receptors	Weight gain, edema, hepatotoxicity
Exenatide	Stimulate insulin secretion, suppress glucagon release	Nausea, vomiting, diarrhea
Nateglinide Repaglinide	Stimulate insulin secretion	Hypoglycemia, weight gain, ear congestion
Glimepiride, Glipizide, Glyburide	Stimulate insulin secretion	Hypoglycemia, skin rashes, itching
Sulphonyl ureas (Glibenclamide)	SU receptors in beta cell	Hypoglycemia, weight gain, skin reactions

Table 1.
Currently used therapeutic agents, their molecular action, and adverse effects.

Flavonoid	Mechanism of Action
Quercetin	Reduce oxidative stress, inhibit alpha-glucosidase, stimulates glucose uptake in muscle cells *via* AMPK pathway, and inhibit advanced glycated end product formation
Kaempferol	Protect beta cells and inhibit alpha amylase
Myricetin	Promote glycogen synthesis and inhibit alpha amylase
Apigenin	Protect beta cells
Luteolin	Inhibit alpha amylase and glucosidase
Genistein	Stimulate insulin secretion, and inhibit alpha amylase
Epicatechin	Protect beta cells
Hesperetin	Inhibit alpha glucosidase, stimulate glucose uptake
Rutin	stimulate glucose uptake Inhibit alpha glucosidase and reduce oxidative stress

Table 2.
Bioflavonoids and their mechanism of action.

pharmacological effects include hepatoprotective, anti-inflammatory, anticancer, hypoglycemic, hypolipidemic, and cardiovascular protection. Additionally, it has been shown in several clinical trials to be advantageous for type 2 diabetes, hyperlipidemia, and NAFLD [12].

2. Natural source of quercetin

Quercetin is widely distributed in the roots, leaf, and fruit of many medicinal plants, including *Mangifera indica, Emblica officinalis, Withania somnifera, Cuscuta reflexa, Santalum album, Curcuma domestica valenton,* and *Foeniculum vulgare*. It is one of the dietary flavonoids used by humans that is now the most common. **Table 3** gives the amount of quercetin present in food sources [13].

Food source	Quercetin in mg/100 g	Food source	Quercetin in mg/100 g
Spinach	27.2	Radicchio	32.0
Black tea	2.0	Fennel leaves	49.0
Red grapes	1.38	Kale	23.0
Capers	233	Cherries	2.7
Onion	22.0	Red apples	4.7
Radish leaves	70.0	Green tea	2.7
Oregano	42.0	Red apples	4.7
Dill	79.0	Green tea	2.7
Chili pepper	32.6	Cilantro	53.0
Lemon	2.29	Coriander	5.00
Tomato	2.7	Radish	70.37
Coen poppy	26.30	Okra	24.00
Pear	0.59	Plum	2.34
Fig	0.87	Strawberry	1.02
Blueberry	9.92	Blackberry	2.70

Table 3.
Food rich in quercetin.

3. Physicochemical properties of quercetin

The melting point of quercetin is 313–314°C, and it is a bright yellow needle-like crystal. It has the chemical formula C15H10O7 with a molecular mass of 302.23. Quercetin is also known as 3,3′,4′,5,7-pentahydroxyflavone or 3,3′,4′,5,7-pentahydroxy-2-phenylchromen-4-one. The flavonol subclass of flavonoids includes quercetin, which has two aromatic rings (A and B) connected by a three-carbon pyrone ring (C). Each of the following locations has a hydroxyl group: 3, 3′, 4′, 5, and 7 as shown in **Figure 2**. Quercetin is soluble in ethanol, methanol, and ethyl acetate, slightly soluble in petroleum ether, benzene, ether, and chloroform, and nearly insoluble in water [14]. It is regarded as having a strong antioxidant effect since its molecular arrangement is acknowledged as the optimum one for scavenging free radicals and binding transition metal ions.

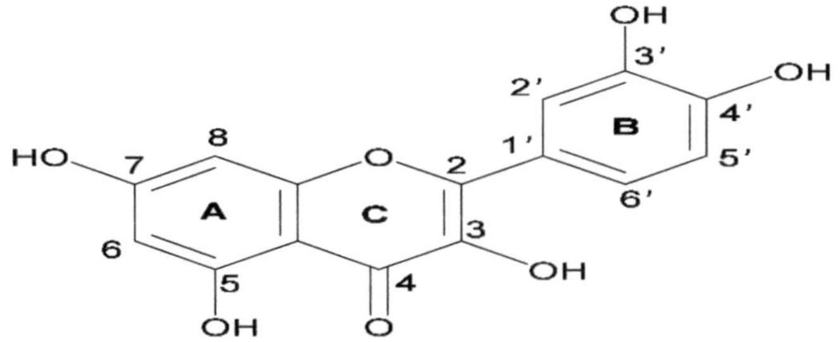

Figure 2.
Quercetin's chemical structure, with (a, B) and (C) standing in for the benzene and pyrone rings, respectively.

4. Bioavailability of quercetin

Fruits and vegetables contain glycoconjugates of quercetin. The quantity and kind of sugar moiety affect the solubility and absorption of glycoconjugates. Quercetin's, *in vivo* metabolism, absorption, and bioavailability, have been thoroughly investigated in both human and animal models. About 65–81% of quercetin aglycone form is hydrolyzed and absorbed in the small intestine. It is then delivered to the liver *via* the portal circulation, where it is digested and bound to albumin in plasma. After 0.7–7 hours of consumption, the plasma level of quercetin reaches its highest [9]. Unabsorbed quercetin is bio-transformed in the intestine by enzymes from the gut microbiota through glucuronidation, hydroxylation, methylation, and sulfonylation. Quercetin is eliminated from humans through their feces or urine. When quercetin (10 mg/70 kg of body weight) was dissolved in grape juice, vegetable homogenate, and white wine, the resulting serum concentrations were 10.8, 25.3, and 12.7 ng/L, respectively [15]. While administration of a 22 mg quercetin capsule produced a plasma concentration of 109 nmol/L, the bioavailability of quercetin is also influenced by the food matrix. Quercetin's bioavailability varies depending on the pH and temperature of the environment. Additionally, quercetin's bioavailability is improved when consumed alongside vitamin C, folate, and other flavonoids [9].

5. Mechanism of action of quercetin against diabetes

5.1 Effects of quercetin on glycemia

Quercetin is a promising drug target for treating diabetes. Several mechanisms, including improved insulin sensitivity, promoted glycogen synthesis, and reduced insulin resistance, have been put up to explain quercetin antihyperglycemic effects. It enhances glucose metabolism and insulin production, which all work together to improve the insulin sensitization effect and also quercetin inhibits the enzymes glucosidase and amylase. In addition, quercetin maintained the mass and functionality of β-cells, hence enhancing the serum insulin effect, and lowered blood glucose in streptozotocin (STZ)-induced diabetes mice. Quercetin, on the other hand, improved the β-cells' ability to secrete insulin and prevented diabetes by reducing oxidative stress in alloxan-diabetic animal models [10].

Quercetin stimulates the GLUT4 expression and endogenous GLUT4 translocation by increasing estrogen receptor, subsequently enhancing phosphorylation of both phosphatidylinositol-3-kinase/Akt (PI3K/Akt) and AMP-activated protein kinase/Akt (AMPK/Akt) signal pathways, increasing the glucose uptake in skeletal muscle cells. Que. acts similarly to rosiglitazone-like drugs such as glycogen phosphorylase (GP) and peroxisome proliferator-activated receptor γ(PPARγ) agonist on insulin-receptor signaling and glucose transport to increase the utilization of glucose [16]. Quercetin might considerably raise plasma insulin levels and lower blood glucose levels by preserving the bulk and activity of cells and enhancing serum insulin action. Additionally, the establishment of the diabetic state in alloxan-diabetic animal models is based on the death of cells brought on by oxidative stress. It has a variety of ways to harm beta cells. However, by lowering oxidative stress in diabetic patients, the quercetin can lessen islet cell failure, improve cell insulin secretion, and further prevent DM.

5.2 Mechanisms of quercetin's hypoglycemic effects

5.2.1 Enhancement of glucose uptake in tissue

In people, quercetin can affect a variety of organs, including the liver, pancreas, small intestine, and muscles. Skeletal muscles that are responsive to insulin are where more than 80% of glucose is absorbed by humans. Therefore, a problem with skeletal muscle glucose absorption might affect the body's overall glucose balance and eventually cause type 2 diabetes. In skeletal muscles, quercetin increases adenosine monophosphate kinase (AMPK), which, in turn, stimulates GLUT4 receptors [17]. Through GLUT4, which facilitates glucose entry into cells, glucose is processed, which controls the glucose level. Additionally, quercetin stimulates AMPK activity in hepatocytes and blocks glucose 6-phosphatase. The main signaling molecules that control cellular GLUT4 expression are AMPK and CaMKII. Exercise is also an effective GLUT4 expression activator, which enhances insulin action and muscle glycogen synthesis (**Figure 3**). On the other hand, quercetin administration decreases intestine sodium-dependent glucose absorption and GLUT2 expression [18]. Quercetin functions similarly to rosiglitazone as a PPAR gamma agonist in that it influences signal transduction and improves the utilization of glucose through affecting glucose transport and insulin-receptor signaling [10].

5.2.2 Release of insulin from β-cells

Pancreatic beta cells secrete insulin, which regulates blood glucose levels when blood sugar levels rise. Reduced viability and functionality of pancreatic cells are frequently linked to diabetic problems. Quercetin improves glucose metabolism and insulin secretion, has a superior hypoglycemic impact, and has an insulin-sensitizing effect by encouraging pancreatic-cell proliferation [19]. In INS-1E (insulin-secreting rat insulinoma) cells, quercetin may help to increase glucose-stimulated insulin

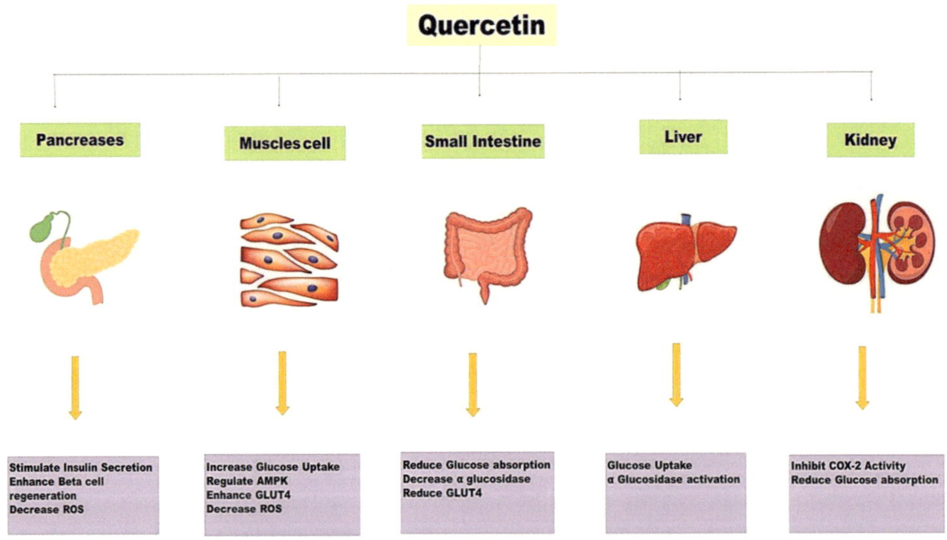

Figure 3.
Mechanism of quercetin as antidiabetic agent.

secretion and insulin expression. According to a study, quercetin can prevent the malfunctioning of pancreatic cells brought on by cholesterol, preserving insulin production triggered by glucose and glycemic management [19]. By specifically activating extracellular signal-regulated kinase 1/2 (ERK1/2), quercetin might greatly improve the insulin secretion of INS-1 pancreatic cells. Moreover, by triggering the intracellular Ca^{2+} signaling system, quercetin dramatically increased the amount of insulin secreted. Additionally, during an oral glucose tolerance test in diabetic rats, quercetin considerably raised the insulin sensitivity index at 30 minutes while significantly lowering the levels of plasma cholesterol, fasting plasma insulin, and postprandial glucose [20]. Regarding this, the beneficial effects of quercetin on islet cells can be divided into three categories, namely an increase in insulin secretion, protection of -cells, and promotion of islet-cell proliferation.

5.2.3 Insulin resistance

Insulin resistance (IR) is the term used to describe the decline in insulin's effectiveness in promoting glucose uptake and utilization, which ultimately results in hyperglycemia. High levels of fatty acids and glucose promote an excessive buildup of ROS, which can damage cells and lead to insulin resistance in peripheral metabolic tissues [10]. In other studies, it was discovered that long-term hyperglycemia decreased the binding of the insulin receptor and the substrate (IRS-1), increased the expression of inducible nitric oxide synthase while decreasing endothelial nitric oxide synthase, and ultimately caused insulin resistance in diabetic rats. Quercetin's effects on glycemic control postulated a number of pathways to account for the advantageous effects of this polyphenol. Inducible nitric oxide synthase expression was raised as a result of the anti-oxidative protective activity on the pancreatic islets of endothelial nitric oxide synthase, which, in turn, caused insulin resistance in diabetic rats. These advantageous effects of quercetin may be attained by decreasing the buildup of cyclic adenosine phosphoric acid (cAMP) and the input of free fatty acids, activating protein kinase A (PKA), maintaining cyclic nucleotide-dependent phosphodiesterase 3B (PDE3B), and increasing two acyl glycerol (DAG) [21].

5.2.4 Alpha-glucosidase inhibition to decrease glucose absorption

One of the therapeutic methods is to suppress the carbohydrate hydrolyzing enzymes, amylases and glucosidase, in the human digestive system in order to delay the absorption of glucose [22]. Consequently, finding glucosidase inhibitors in plants has become a very important endeavor. The intestinal GLUT2 enzyme is severely inhibited by quercetin, which lowers the absorption of glucose [23]. According to numerous reports, quercetin inhibits yeast glucosidases more potently than acarbose. Inhibiting amylase and glucosidases may, therefore, be useful in lowering blood glucose levels after consuming a diet high in mixed carbohydrates. To find out the quercetin and its glycoside derivatives' inhibitory effects on rat intestine glucosidases in an *in vitro* experiment. The findings revealed that quercetin had a high ORAC value, reduced glucosidase activity, and may have physiological benefits for the treatment of diabetes despite the need for additional *in vivo* research. Diabetes patients experience hyperglycemia following a meal as a result of elevated glucosidase and amylase activity. Inhibiting amylase and glucosidases may, therefore, be useful in lowering blood glucose levels following the administration of a mixed carbohydrate diet. In conclusion, reducing insulin resistance, influencing glucose metabolism, and safeguarding islet cells all contribute to the mechanism of quercetin hypoglycemic activity (**Figure 3**) [24].

6. Effects of quercetin on diabetic complications

6.1 Effect on diabetic liver disorders

The term "nonalcoholic fatty liver disease" (NAFLD) refers to a clinic pathological syndrome with excessive intracellular fat deposition that is linked to metabolic illnesses such as hyperlipidemia, insulin resistance, and type 2 diabetes [25]. A growing body of research indicates a clear connection between diabetes and the emergence of liver problems. Diabetes has been linked to an increase in CYP2E1 in the liver, and the suppression of the enzyme primarily protects the liver from oxidative damage. It was discovered that STZ-induced diabetes in rats resulted in hyperglycemia, body weight loss, altered hepatocyte ultrastructure, elevated protein levels, and enhanced CYP2E1 activity in the liver [26]. In STZ-induced diabetic rats, the impact of quercetin on liver apoptosis was examined. The findings demonstrated that 15 mg/kg (IP) quercetin had a strong protective effect against liver cell damage by lowering the number of apoptotic cells; hence, quercetin can be a useful drug for the treatment of diabetes in rats. On experimentally developed diabetic rats, quercetin produced hepatoprotective effects *via* the anti-apoptosis action. Since oxidative stress is improved and hepatocyte apoptosis is inhibited, differing doses of quercetin may be effective in avoiding diabetic liver damage in part because of these effects [25, 27].

6.2 Effect on diabetic retinopathy

The involved mechanisms in diabetes-induced diabetic retinopathy mainly include oxidative stress, inflammation, neurodegeneration, and damaged retinal vasculature. Diabetes-related alterations to the retina were significantly prevented by quercetin therapy [28]. Notably, the protective effect of quercetin was significantly reduced by concomitant HO-1 inhibition. These findings suggest that HO-1 is essential for quercetin's ability to protect the nervous system in DR. Quercetin inhibited high glucose-induced cell proliferation by reducing vascular endothelial growth factor (VEGF) production, according to an in vitro experiment using human retinal endothelial cells. Quercetin also had a neuroprotective effect against diabetic retinopathy [29]. Inhibition of the HMGB1/TLR4/NF-B/NLRP3 inflammation/IL-1/IL-18 axis, suppression of VEGF and sICAM-1 secretion, and encouragement of BDNF secretion *via* HO-1 overexpression make up the underlying mechanism. Quercetin protects against neuronal damage in diabetic retina, perhaps by increasing levels of neurotrophic factors and preventing neuronal death. In order to stop neurodegeneration in DR, quercetin may, therefore, be an appropriate therapeutic agent [30].

6.3 Effect on diabetic nephropathy

One of the main microvascular consequences of diabetes, diabetic nephropathy (DN), is marked by high incidence, low diagnosis rate, lengthy duration, expensive treatment, and high rates of disability and fatality. End-stage renal failure may result from mesangial enlargement and thickening of basement membranes, which cause pathological alterations such as glomerulosclerosis, interstitial fibrosis, and tubular atrophy [31]. Numerous factors, including abnormalities in lipid synthesis, oxidative stress, changes in renal hemodynamics, and signaling pathways for polyol and mitogen-activated protein kinase, all have a role in the outcome and progression of DN. Through the inhibition of protein kinase C (PKC) activity, downregulation of TGF-1 expression, reduction of extracellular matrix production, and delay of renal hypertrophy, quercetin may help the kidney shape that has been affected by hyperglycemia. In addition, quercetin can slow the onset of diabetic nephropathy, prevent degenerative alterations to the kidneys, and enhance glycolipid metabolism in type 2 diabetic rats [32]. Increased insulin resistance and high blood glucose levels have been associated to the release of pro-inflammatory mediators such as IL-1, IL-6, IL-8, IL-4, TNF-, and histamine in brown adipose tissue (**Figure 3**). These mediators are inhibited by quercetin, which also reduces oxidative stress. Quercetin decreases blood glucose reabsorption by lowering DPP-IV and cyclooxygenase-2 (COX-2) activities in the kidneys (**Figure 3**). Quercetin prevented renal development in diabetic rats with kidney disease by stimulating AMPK phosphorylation and decreasing sterol regulatory element binding protein (SREBP)-1c in the kidney [33]. This decreased the tissue dyslipidemia caused by lipid deposits. Quercetin might reduce the oxidative damage brought on by diabetes in the kidney tissues. Additionally, several investigations revealed that diabetic rats with kidney dysfunction had high lipid accumulation. In experimental type 1 diabetic rats, quercetin prevented renal progression by phosphorylating the 5′ adenosine monophosphate (AMP)-activated protein kinase (AMPK) and suppressing sterol regulatory element binding protein (SREBP)-1c in the kidney [32]. This reduced the tissue dyslipidemia condition. As a result, quercetin may 1 day be used as a clinical therapy intervention to stop DM's neurodegenerative consequences.

6.4 Effect on diabetic reproductive disorders

Male diabetic patients frequently experience clinical issues with diabetes-induced reproductive dysfunction, which negatively impacts their sexual lives and ability to conceive. Within five to ten years of the disease's beginning, the reproductive function declined in the majority of diabetic patients. The positive effects of quercetin on diabetes-induced reproductive dysfunction processes, such as sexual dysfunction, testicular dysfunction, or infertility, have been demonstrated in animal trials, but the underlying mechanisms are still unknown [33]. In adult male STZ-induced diabetic SD rats, a recent study examined the effects of quercetin on male sexual behavior and sperm quantity. According to additional research, treating damaged rat testes with an appropriate dosage of quercetin led to a considerable rise in the testis and epididymis weights, SOD activity, and levels of sexual hormones. On the impaired reproductive system of diabetics, quercetin has protective properties [34, 35]. Part of the protective impact of quercetin on sexual behavior, testicular cell damage, and spermatogenic damage brought on by diabetes can be attributed to anti-oxidation, cell proliferation, and anti-apoptosis.

6.5 Effect on diabetic neurodegenerative and neuroprotective disease

According to prior studies, chronic hyperglycemia, oxidative stress, and cholinergic dysfunction are three simple ways to cause diabetic neurodegenerative diseases, which primarily include cognitive and memory deficits, Alzheimer's disease (AD), stroke, and Parkinson's disease [10]. These diseases are a result of changes in the central nervous system. Numerous research on animals has demonstrated the therapeutic effects of antidiabetics, antioxidants, and acetylcholine esterase (AChE) inhibitors on diabetic neurodegenerative disorders, particularly memory loss and cognitive dysfunction. In many *in vitro* and *in vivo* scenarios, quercetin exhibits neuroprotective or increased neurogenesis effects; however, the underlying processes are still not completely understood [36]. In the neurological system, quercetin can guard against neuronal loss or injury brought on by glutamate excitotoxicity, I/R A peptide, and other neurotoxic stimuli. Quercetin significantly reduced cerebral blood flow (CBF) and blood glucose levels, avoided memory loss, enhanced antioxidant enzyme activity, and mitigated abnormalities of the cholinergic system, as well as the consumption of brain energy. Diets high in quercetin should be promoted in order to ward off diabetic vascular and neurological disorders [37]. According to research, people with diabetes, particularly type 2 diabetes, frequently have reduced levels of BDNF, which may be a factor in the disease's cognitive deficits and other consequences. BDNF is well known for its ability to protect the brain. It encourages the development of new neurons and supports the survival and functionality of existing ones. In the context of diabetes, maintaining appropriate BDNF levels may be crucial for shielding nerve cells from harm brought on by oxidative stress and high blood sugar levels [38].

Quercetin supplementation may raise BDNF levels, which may help with cognitive function and memory, both of which can be affected by diabetes-related diseases, such diabetic neuropathy. Furthermore, a western blot study revealed that quercetin enhanced the levels of brain-derived neurotrophic factor (BDNF), nerve growth factor (NGF), postsynaptic density 93 (PSD93), and postsynaptic density 95 (PSD95) in the brain of db/db mice. The pro-inflammatory cytokines IL-1 and IL-18, as well as NLRP3, ASC, and cleaved Caspase-1, were all expressed less, while SIRT1 was expressed more in response to quercetin. It also enhanced the protein expression of SIRT1 and lowered the expression of NLRP3 inflammation-related proteins. The present findings suggest that the SIRT1/NLRP3 pathway may be an important mechanism underlying quercetin's neuroprotective action against DE [39].

6.6 Quercetin and other diabetic complication

One of the most challenging types of pain to manage is diabetic neuropathic pain, a significant micro-vascular consequence of diabetes mellitus (DM). There are not many reports on the use of quercetin to treat diabetic neuropathic pain. The antinociceptive effects of quercetin in normal mice and animals with diabetes caused by STZ. Diabetes patients frequently experience depression, which impairs glucose control and raises the risk of diabetic complications. Clinical care for diabetics has come to be seen as increasingly dependent on the accurate diagnosis and treatment of concomitant depression [40]. Malignant diabetic cardiomyopathy has a high incidence and fatality rate. This type of cardiomyopathy causes hypertension and coronary artery disease and is linked to mitochondrion abnormalities, hyperinsulinemia, insulin resistance, and endoplasmic reticulum modifications. Chronic hyperglycemia causes oxidative stress, which, in turn, causes issues with the central nervous system and may cause neurodegenerative diseases including Parkinson's and Alzheimer's. Recent research has demonstrated that quercetin treatment alleviated memory impairment and decreased brain energy metabolism in rats with STZ-induced memory impairment by lowering ATP content in a dose-dependent manner [41]. The risk of macrovascular consequences, such as hypertension, cardiomyopathy, and coronary artery disorders, is increased by persistent insulin resistance and hyperglycemia. Quercetin, with or without glibenclamide, was observed to lessen cardiomyopathy-related damage in STZ-induced rats in a dose-dependent manner. Recent research has shown that quercetin enhanced cardioprotection in STZ-induced mice by boosting endothelial cell receptors and nitric oxide generation. Another investigation including type 2 diabetes women found quercetin supplements to significantly lower systolic blood pressure. Quercetin may be an effective preventative measure against cardiovascular problems in diabetics; however, more thorough investigation and research are needed [41].

7. Conclusion and future Prospect

A common natural flavonol substance is quercetin. It can be obtained from common fruits and vegetables, as well as some common medicinal plants. The pharmacological effects and molecular mechanisms of quercetin in the management of diabetes were thoroughly reviewed in this chapter. Future preclinical and clinical research on quercetin may benefit from using the material as a resource.

Quercetin has a wide range of pharmacological activities, including actions that are antihyperglycemic, according to studies. It helps reduce hyperlipidemia and hyperglycemia. Through a number of molecular mechanisms, including AMPK, which controls GLUT4 expression in adipose tissue and muscles, quercetin lowers blood glucose levels, increases glucose tolerance, and improves pancreatic beta-cell function. Quercetin has been demonstrated to enhance pancreatic-cell proliferation, insulin sensitivity, glucose metabolism, and insulin production in diabetic animal models. Quercetin has been found to be quite important in the management of diabetes because of its many advantages. Numerous investigations are being conducted to ascertain whether quercetin has the potential to become a future antidiabetic drug. These would provide a safer alternative to the present synthetic medications on the market, which have negative side effects. To learn more about the benefits of quercetin on reducing blood glucose levels and boosting insulin release in T2DM, including the molecular basis of these actions, additional research—from animal models to human trials—is necessary.

In conclusion, more research is needed to determine the precise mechanism of QE on various disorders. QE may be used to successfully prevent some of the clinical consequences of diabetes according to the findings reported in the current review. Future research will focus on QE's involvement in the management of diabetes and its negative effects.

Author details

Priya Mijgar* and Uday Deokate
Government College of Pharmacy, Aurangabad, Maharashtra, India

*Address all correspondence to: pvmijgar.yipb@gmail.com

References

[1] Sok Yen F, Shu Qin C, Tan Shi Xuan S, Jia Ying P, Yi Le H, Darmarajan T, et al. Hypoglycemic effects of plant flavonoids: A review. Evidence-based Complementary and Alternative Medicine. 2021;**2021**:2057333

[2] Koye DN, Magliano DJ, Nelson RG, Pavkov ME. The global epidemiology of diabetes and kidney disease. Advances in Chronic Kidney Disease. 2018;**25**(2):121-132

[3] Samarghandian S, Azimi-Nezhad M, Samini F, Farkhondeh T. Chrysin treatment improves diabetes and its complications in liver, brain, and pancreas in streptozotocin-induced diabetic rats. Canadian Journal of Physiology and Pharmacology. 2016;**94**(4):388-393

[4] Kawser Hossain M, Abdal Dayem A, Han J, Kumar Saha S, Yang GM, Choi HY, et al. Recent advances in disease modeling and drug discovery for diabetes mellitus using induced pluripotent stem cells. International Journal of Molecular Sciences. 2016;**17**(2):256

[5] Coman C, Rugina OD, Socaciu C. Plants and natural compounds with antidiabetic action. Notulae Botanicae Horti Agrobotanici Cluj-Napoca. 2012;**40**(1):314-325

[6] American Diabetes Association. Diagnosis and classification of diabetes mellitus. Diabetes Care. 2014;**37** (Suppl. 1):S81-S90

[7] Pereira DF, Cazarolli LH, Lavado C, Mengatto V, Figueiredo MSRB, Guedes A, et al. Effects of flavonoids on α-glucosidase activity: Potential targets for glucose homeostasis. Nutrition. 2011;**27**(11-12):1161-1167

[8] Ansari P, Choudhury ST, Seidel V, Rahman AB, Aziz MA, Richi AE, et al. Therapeutic potential of quercetin in the management of type-2 diabetes mellitus. Life. 2022;**12**(8):1146

[9] Dhanya R. Quercetin for managing type 2 diabetes and its complications, an insight into multitarget therapy. Biomedicine & Pharmacotherapy. 2022;**146**:112560

[10] Shi GJ, Li Y, Cao QH, Wu HX, Tang XY, Gao XH, et al. In vitro and in vivo evidence that quercetin protects against diabetes and its complications: A systematic review of the literature. Biomedicine & Pharmacotherapy. 2019;**109**:1085-1099

[11] Maalik A, Khan FA, Mumtaz A, Mehmood A, Azhar S, Atif M, et al. Pharmacological applications of quercetin and its derivatives: A short review. Tropical Journal of Pharmaceutical Research. 2014;**13**(9):1561-1566

[12] Prysyazhnyuk V, Voloshyn O. Effects of comprehensive treatment with quercetin administration on biochemical blood parameters and pro-and anti-inflammatory cytokines in nonalcoholic fatty liver disease patients. The Pharma Innovation Journal. 2017;**6**:386-389

[13] Azeem M, Hanif M, Mahmood K, Ameer N, Chughtai FRS, Abid U. An insight into anticancer, antioxidant, antimicrobial, antidiabetic and anti-inflammatory effects of quercetin: A review. Polymer Bulletin. 2023;**80**(1):241-262

[14] Yi H, Peng H, Wu X, Xu X, Kuang T, Zhang J, et al. The therapeutic effects and mechanisms of quercetin on

metabolic diseases: Pharmacological data and clinical evidence. Oxidative Medicine and Cellular Longevity. 2021;**2021**:6678662

[15] Goldberg DM, Yan J, Soleas GJ. Absorption of three wine-related polyphenols in three different matrices by healthy subjects. Clinical Biochemistry. 2003;**36**(1):79-87

[16] Srinivasan P, Vijayakumar S, Kothandaraman S, Palani M. Anti-diabetic activity of quercetin extracted from Phyllanthus emblica L. fruit: In silico and *in vivo* approaches. Journal of Pharmaceutical Analysis. 2018;**8**(2):109-118

[17] Mohammad A, Sharma V, McNeill JH. Vanadium increases GLUT4 in diabetic rat skeletal muscle. Molecular and Cellular Biochemistry. 2002;**233**:139-143

[18] Eid HM, Martineau LC, Saleem A, Muhammad A, Vallerand D, Benhaddou-Andaloussi A, et al. Stimulation of AMP-activated protein kinase and enhancement of basal glucose uptake in muscle cells by quercetin and quercetin glycosides, active principles of the antidiabetic medicinal plant vaccinium vitis-idaea. Molecular Nutrition & Food Research. 2010;**54**(7):991-1003

[19] Li JM, Wang W, Fan CY, Wang MX, Zhang X, Hu QH, et al. Quercetin preserves β-cell mass and function in fructose-induced hyperinsulinemia through modulating pancreatic Akt/FoxO1 activation. Evidence-based Complementary and Alternative Medicine. 2013;**2013**:303902

[20] Adewole SO, Caxton-Martins EA, Ojewole JA. Protective effect of quercetin on the morphology of pancreatic β-cells of streptozotocin-treated diabetic rats. African Journal of Traditional, Complementary and Alternative Medicines. 2007;**4**(1):64-74

[21] Choi HN, Jeong SM, Huh GH, Kim JI. Quercetin ameliorates insulin sensitivity and liver steatosis partly by increasing adiponectin expression in ob/ob mice. Food Science and Biotechnology. 2015;**24**:273-279

[22] Jo SH, Ka EH, Lee HS. Comparison of antioxidant potential and rat intestinal α-glucosidases inhibitory activities of quercetin, rutin, and isoquercetin. International Journal of Applied Research in Natural Products. 2009;**2**:52-60

[23] Soltesova-Prnova M, Milackova I, Stefek M. 3′-O-(3-Chloropivaloyl) quercetin, α-glucosidase inhibitor with multi-targeted therapeutic potential in relation to diabetic complications. Chemical Papers. 2016;**70**(11):1439-1444

[24] Kim JH, Kang MJ, Choi HN, Jeong SM, Lee YM, Kim JI. Quercetin attenuates fasting and postprandial hyperglycemia in animal models of diabetes mellitus. Nutrition Research and Practice. 2011;**5**(2):107-111

[25] Agarwal N, Sharma AK, Agarwal P, Sharma MK, Awasthi N, Mishra A, et al. Evaluation of kidney and liver disorders In type 2 diabetic patients using USG. Evaluation. 2015;**21**(2):265-271.26

[26] Dias AS, Porawski M, Alonso M, Marroni N, Collado PS, Gonzalez-Gallego J. Quercetin decreases oxidative stress, NF-κB activation, and iNOS overexpression in liver of streptozotocin-induced diabetic rats. The Journal of Nutrition. 2005;**135**(10):2299-2304

[27] Sirovina D, Oršolić N, Končić MZ, Kovačević G, Benković V, Gregorović G.

Quercetin vs chrysin: Effect on liver histopathology in diabetic mice. Human & Experimental Toxicology. 2013;**32**(10):1058-1066

[28] Elbe H, Vardi NİGAR, Esrefoglu MUKADDES, Ates B, Yologlu S, Taskapan C. Amelioration of streptozotocin-induced diabetic nephropathy by melatonin, quercetin, and resveratrol in rats. Human & Experimental Toxicology. 2015;**34**(1):100-113

[29] Thomas AA, Feng B, Chakrabarti S. ANRIL: A regulator of VEGF in diabetic retinopathy. Investigative Ophthalmology & Visual Science. 2017;**58**(1):470-480

[30] Kumar B, Gupta SK, Srinivasan BP, Nag TC, Srivastava S, Saxena R, et al. Hesperetin rescues retinal oxidative stress, neuroinflammation and apoptosis in diabetic rats. Microvascular Research. 2013;**87**:65-74

[31] Huang J, Huang K, Lan T, Xie X, Shen X, Liu P, et al. Curcumin ameliorates diabetic nephropathy by inhibiting the activation of the SphK1-S1P signaling pathway. Molecular and Cellular Endocrinology. 2013;**365**(2):231-240

[32] Sonawane RD, Vishwakarma SL, Lakshmi S, Rajani M, Padh H, Goyal RK. Amelioration of STZ-induced type 1 diabetic nephropathy by aqueous extract of Enicostemma littorale Blume and swertiamarin in rats. Molecular and Cellular Biochemistry. 2010;**340**:1-6

[33] Haneda M, Koya D, Kikkawa R. Cellular mechanisms in the development and progression of diabetic nephropathy: Activation of the DAG-PKC-ERK pathway. American Journal of Kidney Diseases. 2001;**38**(4):S178-S181

[34] Nelli GB, Anand Solomon K, Kilari EK. Antidiabetic effect of α-mangostin and its protective role in sexual dysfunction of streptozotocin induced diabetic male rats. Systems Biology in Reproductive Medicine. 2013;**59**(6):319-328

[35] Niwas Jangir R, Chand Jain G. Diabetes mellitus induced impairment of male reproductive functions: A review. Current Diabetes Reviews. 2014;**10**(3):147-157

[36] Ghosh S, Banerjee S, Sil PC. The beneficial role of curcumin on inflammation, diabetes and neurodegenerative disease: A recent update. Food and Chemical Toxicology. 2015;**83**:111-124

[37] Peeyush KT, Gireesh G, Jobin M, Paulose CS. Neuroprotective role of curcumin in the cerebellum of streptozotocin-induced diabetic rats. Life Sciences. 2009;**85**(19-20):704-710

[38] Costa LG, Garrick JM, Roquè PJ, Pellacani C. Mechanisms of neuroprotection by quercetin: Counteracting oxidative stress and more. Oxidative Medicine and Cellular Longevity. 2016;**2016**:2986796

[39] Hu T, Lu XY, Shi JJ, Liu XQ, Chen QB, Wang Q, et al. Quercetin protects against diabetic encephalopathy via SIRT1/NLRP3 pathway in db/db mice. Journal of Cellular and Molecular Medicine. 2020;**24**(6):3449-3459

[40] Kandhare AD, Raygude KS, Kumar VS, Rajmane AR, Visnagri A, Ghule AE, et al. Ameliorative effects quercetin against impaired motor nerve function, inflammatory mediators and apoptosis in neonatal streptozotocin-induced diabetic neuropathy in rats.

Biomedicine & Aging Pathology. 2012;**2**(4):173-186

[41] Li C, Zhao X, Zheng H, Cai F. GW28-e0635 quercetin retards progression of diabetic cardiomyopathy through modulations of SIRT1 and AMP-activated protein kinase. Journal of the American College of Cardiology. 2017;**70**(16S):C63-C64

Chapter 7

Quercetin: A Potential Drug Candidate for Inflammatory Bowel Disease

Mingrui Li and Yun Gao

Abstract

Over the past decades, the incidence rate of inflammatory bowel disease (IBD) has significantly risen all over the world. Most of the patients with IBD suffer from severe symptoms and complications. Being an autoimmune disease, recent research indicates that certain factors, such as environmental changes, disturbances in intestinal microbiota, abnormal immune responses, and genetic susceptibility, play a role in the pathogenesis of IBD. Nevertheless, the precise cause of IBD remains ambiguous. Therefore, there is no known cure for IBD. Moreover, traditional medications have troublesome side effects. For these reasons, some phytochemicals with more tolerance and less adverse effects capture the interest of medical scientists. Flavonoid, a natural anti-inflammatory compound, has recently been validated for its efficacy in IBD treatment. Among the extensive flavonoid family, comprising over 5000 members, quercetin has emerged as a promising drug candidate for treating IBD, supported by substantial preclinical evidence. Currently, quercetin participates in regulating IBD through several pathways, such as antioxidant properties, improvement of the intestinal barrier, modulation of the microbiota, immune response, and regulation of the enteroendocrine system in the gut. In brief, quercetin, a natural compound with anti-inflammatory activity, demonstrates a huge potential as a candidate drug for IBD treatment.

Keywords: flavonoid, quercetin, phytochemical, therapy, inflammatory bowel disease

1. Introduction

Inflammatory bowel disease (IBD) encompasses Crohn's disease (CD), ulcerative colitis (UC), as well as IBD unclassified (IBDU). IBD is a set of chronic or recurrent inflammatory conditions, mainly affecting the small and large intestine [1]. The most observed symptoms of IBD are abdominal pain, unexplained fever, fatigue, diarrhea, and loss of appetite, as well as weight loss [2]. UC manifests as persistent inflammation predominantly in the colon, whereas CD exhibits localized inflammation across the entire gastrointestinal tract. Furthermore, inflammation in UC is restricted to one layer, namely the colonic mucosa. Contrastingly, CD can affect two or multiple layers of the intestinal wall [3]. Presently, the global prevalence of IBD is on the rise. Projections suggest that the population afflicted with IBD in the Western world could surpass 10 million by 2030 [4]. In addition, the prevalence of IBD in Asia has significantly increased, leading to substantial economic burdens on individuals and societies [5]. To date, studies have shown that its pathogenesis is associated with environmental factors, immunity, microbiota, and genetics. However, the etiology of IBD is still a mystery [6]. Therefore, IBD is not possible to cure. At present, IBD treatment involves drug therapy and surgery. The medications have troublesome side effects with long-term use. In contrast, surgery does not cure CD, and the benefits are usually temporary [7].

Quercetin is a member of the flavonoid family, widely distributed in plants. Quercetin (≥99.5%) has been granted Generally Recognized As Safe (GRAS) status by the U.S. Food and Drug Administration (FDA) and is permitted for incorporation into established food additives in Japan and Korea. For the past few years, the medicinal significance of quercetin has increasingly garnered attention. Like other flavonoids, quercetin possesses antioxidant, anti-tumor, anti-diabetic, anti-hypertensive, and anti-depressant properties [8, 9]. Recent studies suggest that several pathways of quercetin participate in the regulation of IBD, comprising antioxidant properties, maintenance of the intestinal barrier, modulation of the microbiota, immune response, and regulation of the enteroendocrine system in the gut [10, 11]. However, these research findings are predominantly based on preclinical evidence. Therefore, the specific mechanisms need further investigation in the human body. In conclusion, the chapter is mainly aimed to illuminate the therapeutic potential of quercetin in IBD treatment.

2. The pathogenesis of IBD

Recent research suggests that the pathogenesis of IBD, as illustrated in **Figure 1**, is mainly associated with four factors, namely environmental factors, gut microbiota, and genetic susceptibility, as well as excess immunity [6]. Regarding environmental influences, it has been reported that a diet abundant in fruits and vegetables has a positive correlation with a reduced morbidity of Crohn's disease [12]. Conversely, the consumption of fast food high in sugar and fat may trigger the initiation of IBD [13]. Additionally, factors such as smoking, medication, oxidative stress, and psychological stress may impact the development of IBD [14–16].

Furthermore, several studies have demonstrated significant distinctions in the gut microbiota between individuals with IBD and healthy counterparts. The microbiota in IBD patients exhibits altered composition, reduced levels of protective bacteria, and increased presence of pathogenic bacteria, as well as disrupted microbial balance. These observed changes may play a role in the progression of IBD [17–19].

The intestinal immune system consists of both innate and adaptive immunity. Among them, innate immunity encompasses the protective barrier function of the intestinal mucosa, antibacterial proteins, gastric acid, and immune cells, as well as innate cytokines and molecules. In contrast, adaptive immunity (T and B cells) is pathogen-specific and is typically activated when the response of the innate immune system to a pathogen is ineffective [20]. Regulating the immune response to the intestinal microbiota is crucial for maintaining a delicate equilibrium between defensive inflammation and immune tolerance. Any disturbance to this equilibrium can lead to abnormal immunity, ultimately initiating the development of IBD [21].

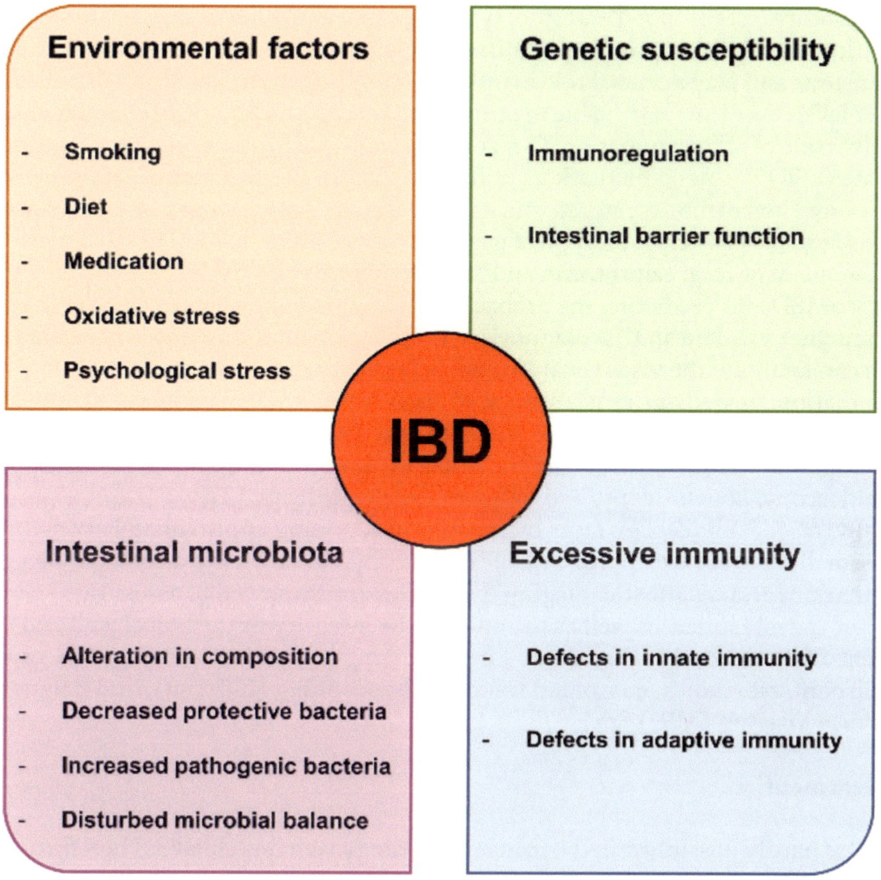

Figure 1.
Possible factors leading to IBD development.

Genetic susceptibility has also occurred in individuals with IBD [22]. Immune regulation and epithelial barrier function are crucial steps for keeping intestinal stability, as revealed by the investigation of genes and genetic loci associated with IBD. An impairment of the epithelial barrier can enable the invasion of microbes, subsequently recognized by the innate immune system. This recognition triggers various responses, including tolerogenic, inflammatory, and restorative responses. These responses are partly mediated by the secretion of extracellular mediators, attracting T or B cells [23]. Nonetheless, the exact pathogenesis of IBD is still uncertain so far [6].

3. Current diagnosis and treatment of IBD

3.1 Diagnosis procedure

Consistent or recrudescent bloody diarrhea, accompanied by abdominal pain and frequent defecation should raise suspicion of IBD, particularly in the case of young patients. To confirm a diagnosis of IBD, an array of tests and procedures is imperative for patients.

Laboratory tests of blood and stool typically serve as the initial diagnostic step. Biomarkers, comprising proteins identified in blood and stool, are integral components of these tests and play a crucial role in inflammation detection. Monitoring specific biomarker levels over time is valuable in optimizing therapeutic interventions and ensuring effective control of inflammation. C-reactive protein (CRP) and erythrocyte sedimentation rate (ESR) are blood biomarkers utilized to discern the presence of inflammation in the body. Concurrently, calprotectin and lactoferrin, both proteins found in stool, serve as fecal biomarkers indicative of gastrointestinal inflammation [24]. Moreover, the assessment of fecal calprotectin and lactoferrin proves valuable in monitoring the activity of IBD and predicting the probability of an impending clinical relapse [25].

To further validate an IBD diagnosis, the operation of endoscopy is necessary, which can facilitate the visual examination of the presence and localization of inflammation, as well as the assessment of its severity, and collect tissue samples via biopsies. Moreover, endoscopic procedures can monitor the therapeutic efficacy. Endoscopies used in IBD diagnosis encompass colonoscopy, upper GI endoscopy, sigmoidoscopy, pouchoscopy, and video capsule endoscopy [24].

Conventional endoscopy faces limitations in accessing approximately two-thirds of the small intestine except for capsule endoscopy. Based on it, various radiologic examinations and diagnostic imaging modalities are employed to assess these segments of the intestines, as well as to appraise the integrity of the bowel wall and adjacent areas external to the bowel. These imaging procedures encompass X-rays, barium contrast studies, computed tomography scanning (CT scan), and magnetic resonance imaging (MRI) [24].

3.2 Treatment

IBD is hardly possibly cured because its etiology remains elusive. Therefore, the main objective of treating IBD is to decrease inflammation level and alleviate the signs and symptoms experienced by patients. In best cases, this may lead not only to relieve symptoms but also to reduce risks of complications (tissue fibrosis, stenosis, fistulas, and colon cancer). Typically, the treatment of IBD includes both medication and surgical interventions.

IBD drug therapy primarily encompasses anti-inflammatory drugs, immune system suppressors, biologics, antibiotics, and other medications, as well as nutritional support. Anti-inflammatory drugs often constitute the initial approach in the treatment of IBD, particularly for cases characterized by mild to moderate severity. This category encompasses aminosalicylates (mesalamine, balsalazide, olsalazine) and corticosteroids. Immune system suppressors involve mercaptopurine, methotrexate, azathioprine, and some oral "small molecules" (tofacitinib, upadacitinib, and ozanimod). These drugs can inhibit the immune response, which discharges inflammation-inducing chemicals into the human body. Once released, these chemicals can cause the impairment of the digestive tract. Biologics are an emerging category of therapy, which are targeted at neutralizing proteins causing inflammation in the body. The current products on the market include infliximab, adalimumab, golimumab, certolizumab, vedolizumab, and ustekinumab, as well as risankizumab. Antibiotics may be utilized in conjunction with other medications when infection is a concern in cases of perianal Crohn's disease. Beyond inflammation control, certain medications, such as anti-diarrheal medications and pain relievers, may alleviate the signs and symptoms of IBD. Additionally, under conditions of compromised nutrient absorption or severe weight loss, vitamins and nutritional supplements are indispensable.

Surgery is the last option when all conservative treatments cannot ameliorate the signs and symptoms of IBD, which involve changes in diet and lifestyle and drug therapy as well as alternative medicine. Notwithstanding, the side effects of surgery bring great inconvenience and burden to patients with UC. Moreover, surgery does not cure CD, and the benefits are usually temporary [7].

4. The association between the intake of dietary quercetin and IBD

As of now, there is no direct evidence supporting the mitigation of IBD through dietary quercetin consumption. However, flavonoid intake plays a positive role in alleviating IBD [26, 27]. Because quercetin belongs to the flavonoid family, the intake of dietary quercetin has been indirectly associated with the amelioration of IBD. Currently, several studies have focused on the impacts of flavonoid consumption on healthy individuals and IBD patients. For instance, Bai et al. reported that the mean total flavonoid intake by U.S. adults was 344.83 ± 9.13 mg/day [28]. Kölbel et al. investigated the correlation between the consumption of dietary flavonoids and the overall severity of IBD. They found that in the average Western diet, the total daily intake of flavonoids typically falls within the range of 200–1000 mg/day, but it decreases to 144 mg/day in individuals with IBD. In addition, a low dietary flavonoid intake has a positive correlation with severe IBD [27]. Furthermore, a randomized trial demonstrated that adhering to the Mediterranean diet, abundant in flavonoids, enhanced clinical scores and reduced inflammatory markers among children and adolescents with mild to moderate active IBD were observed [26].

5. Mechanisms of quercetin regulating IBD

Galsanov et al. reported the potential of quercitrin (a glycoside form of quercetin) as an agent to ameliorate intestinal inflammation in 1976 [29]. Following this discovery, other flavonoids have gradually been recognized for their anti-inflammatory properties in IBD. Notably, quercetin stands out as the first identified flavonoid capable of mitigating intestinal inflammation. Currently, several pathways of quercetin regulating IBD have been identified. As depicted in **Figure 2**, these include antioxidant activity, preservation of the epithelial barrier, the regulation of microbiota, and immune system in the gut, as well as the modulation of the enteroendocrine system (**Figure 2**).

5.1 Antioxidant property

Oxidative stress is recognized as a factor that exacerbate symptoms of IBD, such as diarrhea and abdominal pain. In addition, it contributes to the development of IBD [30]. In this process, reactive nitrogen species (RNS) and reactive oxygen species (ROS) have emerged as crucial contributors to the pathogenesis of IBD [31]. Recent research has revealed high levels of RNS and ROS in the inflamed intestine [32]. In IBD patients, excessive mononuclear cells and neutrophils aggregate in the inflamed intestinal tissue followed by increased myeloperoxidase (MPO) level and activated nicotinamide adenine dinucleotide phosphate (NADPH) oxidase system, resulting in the overproduction of superoxide and hypochlorous acid, which have direct cytotoxicity to the affected tissue [33]. Additionally, this pathway also generates a

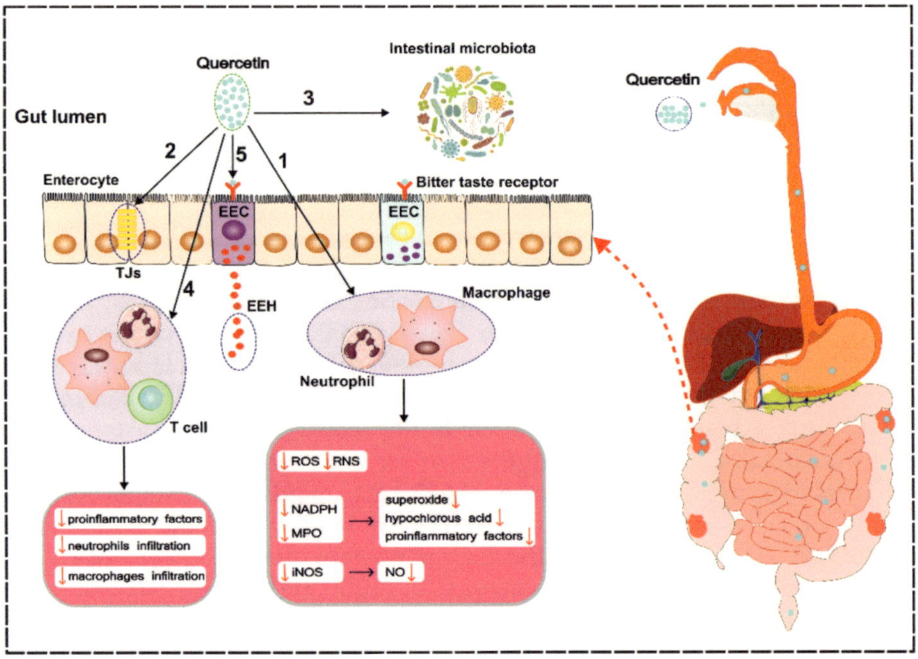

Figure 2.
Mechanisms of quercetin regulating IBD: (1) Antioxidant activity. (2) Preservation of the impaired epithelial barrier. (3) Shaping the intestinal microbiota. (4) Immunomodulatory function. (5) Modulation of the enteroendocrine system (RNS-reactive nitrogen species, ROS-reactive oxygen species, MPO-myeloperoxidase, NADPH-nicotinamide adenine dinucleotide phosphate, NO-nitric oxide, iNOS-inducible nitric oxide synthase, Tjs-tight junctions, EEH-enteroendocrine hormones, EEC-enteroendocrine cell).

considerable number of pro-inflammatory factors [34]. Nonetheless, quercetin can significantly decrease MPO levels and inhibit ROS overproduction in colitis models, thereby blocking the following tissue damage [30, 35]. In normal physiological circumstances, a small amount of nitric oxide (NO) provides protection against inflammatory injury. On the contrary, in the inflammatory state, the overexpression of inducible nitric oxide synthase (iNOS) results in excessive NO production, causing an overproduction of peroxynitrites and followed by intestinal destruction under the action of superoxide anions [36]. Recent studies have shown that quercetin can curb the expression of iNOS, leading to a noticeable decrease in NO production [37, 38]. Therefore, quercetin can decrease the NO-induced enterotoxicity.

5.2 Preserving epithelial barrier

The epithelial barrier in the gut is an essential interface, dividing the external environment and the host tissue and playing a vital role in maintaining intestinal homeostasis [39]. While the loss of the epithelial barrier is a common event in the development of IBD, preserving the integrity of the intestinal epithelial barrier is a critical step in preventing the progression of IBD [40]. Studies have shown that quercetin may have a beneficial effect on the epithelial barrier [41]. For instance, Riemschneider et al. established murine colitis models induced by dextran sulfate sodium (DSS) and found that oral administration of quercetin restored the loss of epithelial integrity by the induction of tight junction proteins [42].

5.3 Shaping microbiota

The dysbiosis of gut microbiota is a vital participant in the development of IBD, characterized by diminished biodiversity, reduced stability, and an elevated presence of *Proteobacteria*, including *Bilophila* and *Enterobacteriaceae*, as well as certain segments of *Bacteroidetes* [43]. Similarly, dysbiosis can induce inflammation in the intestine by disrupting host homeostasis [44]. Recent studies have demonstrated that dietary quercetin has the potential to protect against IBD by improving intestinal dysbiosis. For instance, Lin et al. utilized dietary quercetin to treat *Citrobacter rodentium*-infected mice and observed a significant reduction in the inflammatory level of colitis induced by *Citrobacter rodentium*. Additionally, quercetin was found to enhance the populations of *Bacteroides*, *Bifidobacterium*, *Lactobacillus*, and *Clostridia* while reducing those of *Fusobacterium* and *Enterococcus*. Thus, the amelioration of colitis in mice is probably attributed to the quercetin's ability to modulate the gut microbiota [45]. In addition, Zhu et al. observed a reduced *Firmicutes* population and an increased *Proteobacteria* population in DSS-induced colitis mice. Quercetin was found to restore these changes, thereby affecting the outcomes of microbiota-associated disorders [35].

5.4 Immunomodulatory function

IBD is closely associated with the aberrant response of the intestinal immune system, which contains the innate immune system comprising dendritic cells, macrophages, and neutrophils, as well as the adaptive immune system including T lymphocytes and B lymphocytes. Quercetin is able to diminish the infiltration of macrophages, neutrophils, and Th17 cells, concurrently promoting an increase in Treg cell proportions [42]. In the lipopolysaccharide (LPS)-stimulated macrophages, quercetin was found to inhibit the mRNA expression of key proinflammatory cytokines, namely tumor necrosis factor-alpha (TNF-α), interleukin-1 beta (IL-1β), and interleukin-6 (IL-6), primarily by interfering with the mitogen-activated protein kinase (MAPK) and nuclear factor-κB (NF-κB) pathways [46]. Lin et al. further elucidated the anti-inflammatory properties of quercetin, revealing its ability to suppress the production of IL-17, TNF-α, and IL-6 (proinflammatory cytokines) while simultaneously boosting the release of IL-10 (anti-inflammatory cytokine) in colon tissues [45]. Specifically, quercetin suppresses TNF-α and IL-6 production by modulating the miR-369-3p/C/EBP-β axis in dendritic cells, thereby showing its anti-inflammatory properties [47]. This consistent effect is also observed in intestinal epithelial cells, wherein the restraint on C/EBP-β signaling by quercetin explains its protective activity against IL-6 production induced by the heat shock response in the intestinal mucosa [48].

5.5 The modulation of the enteroendocrine system

Currently, there is a lack of evidence supporting the enteroendocrine regulation of quercetin in IBD. However, flavonoids have potential protective effects against IBD through the enteroendocrine pathway [11]. Accordingly, we conclude that quercetin may also protect against IBD via enteroendocrine regulation.

The mechanisms of flavonoids regulating the enteroendocrine system are as follows: flavonoids can stimulate the secretion of cholecystokinin (CCK), ghrelin, glucagon-like peptide (GLP-1), and glucagon-like peptide (GLP-2) [49–52]. GLP-1/2; exhibit considerable potential in treating IBD due to their ability to promote the restoration of the impaired epithelial barrier; regulate T lymphocyte differentiation and function; and modulate innate immune cells, such as macrophages and dendritic cells [53–56]. Ghrelin has also been shown to mitigate intestinal inflammation in colitis mice [57]. In addition, CCK has demonstrated anti-inflammatory effects [58]. For these reasons, it can be inferred that flavonoids exhibit therapeutic effects on IBD through the enteroendocrine approach.

6. Clinical application of quercetin in IBD

Bioavailability is defined as "the speed and degree to which the active component of a drug is absorbed and become accessible at the site of action", which are crucial factors affecting the efficacy of bioactive compounds in organisms. Similar to other flavonoids, quercetin demonstrates low bioavailability, thereby contributing to reduced efficacy in vivo [59]. Accordingly, the clinical application of quercetin in IBD is unsatisfactory. It has been reported that the bioavailability can be affected by several factors, including glycosylation, molecular weight, and metabolic conversion, as well as interaction with colonic microbiota [60]. Instability, poor absorption and solubility, rapid metabolism, and systemic elimination may result from the above confounding factors, which cause low bioavailability of quercetin and other flavonoids. Consequently, their delivery efficiency and therapeutic efficacy in vivo are significantly decreased.

To enhance the in vivo bioavailability of quercetin, researchers have directed their attention toward optimizing various metabolic processes associated with bioavailability, such as augmenting intestinal absorption [61], improving metabolic stability [62], and modifying the site of absorption [63], among others. In recent years, nano-delivery systems have been utilized for quercetin to accomplish the above objective [64, 65]. For instance, Wang et al. formulated novel quercetin-loaded nanoparticles (NPs) using sodium alginate and N-succinyl chitosan. Their findings indicate that these NPs show enhanced therapeutic efficacy in alleviating DSS-induced colitis in mice, compared with free quercetin [66]. In addition, Jing et al. developed a colon-targeted quercetin delivery system, of which the pectin/Ca^{2+} microspheres were prepared and then crosslinked with oligochitosan. The therapeutic outcomes observed in vivo study indicated that these quercetin microspheres could alleviate symptoms of colitis and preserve both the colon length and the intestinal barrier integrity [67].

7. Conclusions

Quercetin is a common flavonol belonging to the flavonoid family. The current preclinical studies show that quercetin has the potential to protect against IBD. The underlying mechanisms include antioxidant activity, improving the intestinal epithelial barrier, modulating the intestinal microbiota, immunomodulatory function, and regulating the enteroendocrine system. Unfortunately, the clinical application of quercetin in IBD is unsatisfactory due to its low bioavailability. In the future, the design of novel material will contribute to improving the bioavailability of quercetin, thereby promoting its therapeutical effect on IBD.

Author contributions

Mingrui Li drafted, edited and visualized this chapter. Yun Gao supervised and revised the chapter.

Conflict of interest

The authors declare no conflict of interest.

Author details

Mingrui Li[1] and Yun Gao[2]*

1 Department of Medicine, Kussmaul Campus for Medical Research, University of Erlangen-Nürnberg, Erlangen, Germany

2 Department of Endocrinology and Metabolism, Diabetic Foot Care Center, West China Hospital, Sichuan, China

*Address all correspondence to: gaoyun@wchscu.cn

References

[1] Nakase H, Uchino M, Shinzaki S, Matsuura M, Matsuoka K, Kobayashi T, et al. Evidence-based clinical practice guidelines for inflammatory bowel disease 2020. Journal of Gastroenterology. 2021;**56**(6):489-526

[2] Hendrickson BA, Gokhale R, Cho JH. Clinical aspects and pathophysiology of inflammatory bowel disease. Clinical Microbiology Reviews. 2002;**15**(1):79-94

[3] Bouma G, Strober W. The immunological and genetic basis of inflammatory bowel disease. Nature Reviews. Immunology. 2003;**3**(7):521-533

[4] Kaplan GG, Windsor JW. The four epidemiological stages in the global evolution of inflammatory bowel disease. Nature Reviews. Gastroenterology & Hepatology. 2021;**18**(1):56-66

[5] Banerjee R, Pal P, Mak JWY, Ng SC. Challenges in the diagnosis and management of inflammatory bowel disease in resource-limited settings in Asia. The Lancet Gastroenterology & Hepatology. 2020;**5**(12):1076-1088

[6] Guan Q. A comprehensive review and update on the pathogenesis of inflammatory bowel disease. Journal of Immunology Research. 2019;**2019**:7247238

[7] (IBD) Ibd: Mayo Clinic 2022. Available online: https://wwwmayoclinicorg/diseases-conditions/inflammatory-bowel-disease/diagnosis-treatment/drc-20353320 [Accessed: December 1, 2023]

[8] Dodda D, Chhajed R, Mishra J. Protective effect of quercetin against acetic acid induced inflammatory bowel disease (IBD) like symptoms in rats: Possible morphological and biochemical alterations. Pharmacological Reports. 2014;**66**(1):169-173

[9] Dodda D, Chhajed R, Mishra J, Padhy M. Targeting oxidative stress attenuates trinitrobenzene sulphonic acid induced inflammatory bowel disease like symptoms in rats: Role of quercetin. Indian Journal Of Pharmacology. 2014;**46**(3):286-291

[10] Hu S, Zhao M, Li W, Wei P, Liu Q, Chen S, et al. Preclinical evidence for quercetin against inflammatory bowel disease: A meta-analysis and systematic review. Inflammopharmacology. 2022;**30**(6):2035-2050

[11] Li M, Weigmann B. A novel pathway of flavonoids protecting against inflammatory bowel disease: Modulating enteroendocrine system. Metabolites. 2022;**12**(1):31-48

[12] Dolan KT, Chang EB. Diet, gut microbes, and the pathogenesis of inflammatory bowel diseases. Molecular Nutrition & Food Research. 2017;**61**(1):10-29

[13] Hibi T, Ogata H. Novel pathophysiological concepts of inflammatory bowel disease. Journal of Gastroenterology. 2006;**41**(1):10-16

[14] Ardizzone S, Bianchi Porro G. Biologic therapy for inflammatory bowel disease. Drugs. 2005;**65**(16):2253-2286

[15] Ho SM, Lewis JD, Mayer EA, Plevy SE, Chuang E, Rappaport SM, et al. Challenges in IBD research: Environmental triggers. Inflammatory Bowel Diseases. 2019;**25**(Suppl. 2):S13-S23

[16] Lakatos PL. Environmental factors affecting inflammatory bowel disease: Have we made progress? Digestive Diseases. 2009;**27**(3):215-225

[17] Glassner KL, Abraham BP, Quigley EMM. The microbiome and inflammatory bowel disease. The Journal of Allergy and Clinical Immunology. 2020;**145**(1):16-27

[18] Nishida A, Inoue R, Inatomi O, Bamba S, Naito Y, Andoh A. Gut microbiota in the pathogenesis of inflammatory bowel disease. Clinical Journal of Gastroenterology. 2018;**11**(1):1-10

[19] Richard ML, Sokol H. The gut mycobiota: Insights into analysis, environmental interactions and role in gastrointestinal diseases. Nature Reviews. Gastroenterology & Hepatology. 2019;**16**(6):331-345

[20] Ince MN, Elliott DE. Immunologic and molecular mechanisms in inflammatory bowel disease. The Surgical Clinics of North America. 2007;**87**(3):681-696

[21] Abraham C, Cho JH. Inflammatory bowel disease. The New England Journal of Medicine. 2009;**361**(21):2066-2078

[22] Annese V. Genetics and epigenetics of IBD. Pharmacological Research. 2020;**159**:104892

[23] Khor B, Gardet A, Xavier RJ. Genetics and pathogenesis of inflammatory bowel disease. Nature. 2011;**474**(7351):307-317

[24] Crohn's & Colitis Foundation. Available online: https:// wwwcrohnscolitisfoundationorg/ what-is-ibd/diagnosing-ibd [Accessed: December 15, 2023]

[25] Grifka J, Schiegl Geb Gotz J, Fenk-Mayer A, Benditz A. Injection treatment for cervical and lumbar syndromes: Special infiltration techniques. Orthopadie (Heidelb). 2023;**46**(1):195-208

[26] El Amrousy D, Elashry H, Salamah A, Maher S, Abd-Elsalam SM, Hasan S. Adherence to the Mediterranean diet improved clinical scores and inflammatory markers in children with active inflammatory bowel disease: A randomized trial. Journal of Inflammation Research. 2022;**15**:2075-2086

[27] Kölbel B, Hamacher S, Hellmich M, Kruis W. Low dietary flavonoid consumption is associated to severe inflammatory bowel disease. Gastro Hep Advances. 2024;**3**(1):31-37

[28] Bai W, Wang C, Ren C. Intakes of total and individual flavonoids by US adults. International Journal of Food Sciences and Nutrition. 2014;**65**(1):9-20

[29] Galsanov Sh B, Tourova AD, Klimenko ED. Effect of quercitrin on structural changes in the large and small intestines in experimental enterocolitis. Biulleten' Eksperimental'noĭ Biologii i Meditsiny. 1976;**81**(5):623-625

[30] Dong Y, Hou Q, Lei J, Wolf PG, Ayansola H, Zhang B. Quercetin alleviates intestinal oxidative damage induced by H(2)O(2) via modulation of GSH: In vitro screening and In vivo evaluation in a colitis model of mice. ACS Omega. 2020;**5**(14):8334-8346

[31] Rezaie A, Parker RD, Abdollahi M. Oxidative stress and pathogenesis of inflammatory bowel disease: An epiphenomenon or the cause? Digestive Diseases and Sciences. 2007;**52**(9):2015-2021

[32] Piechota-Polanczyk A, Fichna J. Review article: The role of oxidative stress in pathogenesis and treatment of inflammatory bowel diseases. Naunyn-Schmiedeberg's Archives of Pharmacology. 2014;**387**(7):605-620

[33] Alzoghaibi MA. Concepts of oxidative stress and antioxidant defense in Crohn's disease. World Journal of Gastroenterology. 2013;**19**(39):6540-6547

[34] Pavlick KP, Laroux FS, Fuseler J, Wolf RE, Gray L, Hoffman J, et al. Role of reactive metabolites of oxygen and nitrogen in inflammatory bowel disease. Free Radical Biology & Medicine. 2002;**33**(3):311-322

[35] Hong Z, Piao M. Effect of quercetin monoglycosides on oxidative stress and gut microbiota diversity in mice with dextran sodium Sulphate-induced colitis. BioMed Research International. 2018;**2018**:8343052

[36] Veljaca M, Lesch CA, Pllana R, Sanchez B, Chan K, Guglietta A. BPC-15 reduces trinitrobenzene sulfonic acid-induced colonic damage in rats. The Journal of Pharmacology and Experimental Therapeutics. 1995;**272**(1):417-422

[37] Khater SI, Lotfy MM, Alandiyjany MN, Alqahtani LS, Zaglool AW, Althobaiti F, et al. Therapeutic potential of quercetin loaded nanoparticles: novel insights in alleviating colitis in an experimental DSS induced colitis model. Biomedicine. 2022;**10**(7):1654-1677

[38] Romero M, Vera B, Galisteo M, Toral M, Galvez J, Perez-Vizcaino F, et al. Protective vascular effects of quercitrin in acute TNBS-colitis in rats: The role of nitric oxide. Food & Function. 2017;**8**(8):2702-2711

[39] Turner JR. Intestinal mucosal barrier function in health and disease. Nature Reviews. Immunology. 2009;**9**(11):799-809

[40] Sommer K, Wiendl M, Muller TM, Heidbreder K, Voskens C, Neurath MF, et al. Intestinal mucosal wound healing and barrier integrity in IBD-crosstalk and trafficking of cellular players. Frontiers in Medicine (Lausanne). 2021;**8**:643973

[41] Suzuki T, Hara H. Role of flavonoids in intestinal tight junction regulation. The Journal of Nutritional Biochemistry. 2011;**22**(5):401-408

[42] Riemschneider S, Hoffmann M, Slanina U, Weber K, Hauschildt S, Lehmann J. Indol-3-carbinol and quercetin ameliorate chronic DSS-induced colitis in C57BL/6 mice by AhR-mediated anti-inflammatory mechanisms. International Journal of Environmental Research and Public Health. 2021;**18**(5):2262-2278

[43] Nishino K, Nishida A, Inoue R, Kawada Y, Ohno M, Sakai S, et al. Analysis of endoscopic brush samples identified mucosa-associated dysbiosis in inflammatory bowel disease. Journal of Gastroenterology. 2018;**53**(1):95-106

[44] Sultan S, El-Mowafy M, Elgaml A, Ahmed TAE, Hassan H, Mottawea W. Metabolic influences of gut microbiota dysbiosis on inflammatory bowel disease. Frontiers in Physiology. 2021;**12**:715506

[45] Lin R, Piao M, Song Y. Dietary quercetin increases colonic microbial diversity and attenuates colitis severity in Citrobacter rodentium-infected mice. Frontiers in Microbiology. 2019;**10**:1092

[46] Cho SY, Park SJ, Kwon MJ, Jeong TS, Bok SH, Choi WY, et al. Quercetin suppresses proinflammatory cytokines

production through MAP kinases andNF-kappaB pathway in lipopolysaccharide-stimulated macrophage. Molecular and Cellular Biochemistry. 2003;**243**(1-2):153-160

[47] Galleggiante V, De Santis S, Liso M, Verna G, Sommella E, Mastronardi M, et al. Quercetin-induced miR-369-3p suppresses chronic inflammatory response targeting C/EBP-beta. Molecular Nutrition & Food Research. 2019;**63**(19):e1801390

[48] Hungness ES, Robb BW, Luo GJ, Pritts TA, Hershko DD, Hasselgren PO. Proteasome inhibitors activate the transcription factors C/EBP-beta and delta in human intestinal epithelial cells. Biochemical and Biophysical Research Communications. 2002;**290**(1):469-474

[49] Al Shukor N, Ravallec R, Van Camp J, Raes K, Smagghe G. Flavonoids stimulate cholecystokinin peptide secretion from the enteroendocrine STC-1 cells. Fitoterapia. 2016;**113**:128-131

[50] Gonzalez-Abuin N, Martinez-Micaelo N, Margalef M, Blay M, Arola-Arnal A, Muguerza B, et al. A grape seed extract increases active glucagon-like peptide-1 levels after an oral glucose load in rats. Food & Function. 2014;**5**(9):2357-2364

[51] Grau-Bove C, Gonzalez-Quilen C, Terra X, Blay MT, Beltran-Debon R, Jorba-Martin R, et al. Effects of flavanols on enteroendocrine secretion. Biomolecules. 2020;**10**(6):844-856

[52] Kato M, Tani T, Terahara N, Tsuda T. The anthocyanin Delphinidin 3-Rutinoside stimulates glucagon-like Peptide-1 secretion in murine GLUTag cell line via the Ca2+/calmodulin-dependent kinase II pathway. PLoS One. 2015;**10**(5):e0126157

[53] Anbazhagan AN, Thaqi M, Priyamvada S, Jayawardena D, Kumar A, Gujral T, et al. GLP-1 nanomedicine alleviates gut inflammation. Nanomedicine. 2017;**13**(2):659-665

[54] Bang-Berthelsen CH, Holm TL, Pyke C, Simonsen L, Sokilde R, Pociot F, et al. GLP-1 induces barrier protective expression in Brunner's glands and regulates colonic inflammation. Inflammatory Bowel Diseases. 2016;**22**(9):2078-2097

[55] Buchman AL, Katz S, Fang JC, Bernstein CN, Abou-Assi SG, Teduglutide Study G. Teduglutide, a novel mucosally active analog of glucagon-like peptide-2 (GLP-2) for the treatment of moderate to severe Crohn's disease. Inflammatory Bowel Diseases. 2010;**16**(6):962-973

[56] Qi KK, Lv JJ, Wu J, Xu ZW. Therapeutic effects of different doses of polyethylene glycosylated porcine glucagon-like peptide-2 on ulcerative colitis in male rats. BMC Gastroenterology. 2017;**17**(1):34

[57] Gonzalez-Rey E, Chorny A, Delgado M. Therapeutic action of ghrelin in a mouse model of colitis. Gastroenterology. 2006;**130**(6):1707-1720

[58] Saia RS, Ribeiro AB, Giusti H. Cholecystokinin modulates the mucosal inflammatory response and prevents the lipopolysaccharide-induced intestinal epithelial barrier dysfunction. Shock. 2020;**53**(2):242-251

[59] Enayat S, Seyma Ceyhan M, Taskoparan B, Stefek M, Banerjee S. CHNQ, a novel 2-Chloro-1,4-naphthoquinone derivative of quercetin, induces oxidative stress and autophagy both in vitro and in vivo. Archives of Biochemistry and Biophysics. 2016;**596**:84-98

[60] Thilakarathna SH, Rupasinghe HP. Flavonoid bioavailability and attempts for bioavailability enhancement. Nutrients. 2013;**5**(9):3367-3387

[61] Shen Q, Li X, Li W, Zhao X. Enhanced intestinal absorption of daidzein by borneol/menthol eutectic mixture and microemulsion. AAPS PharmSciTech. 2011;**12**(4):1044-1049

[62] Walle T. Methylation of dietary flavones greatly improves their hepatic metabolic stability and intestinal absorption. Molecular Pharmaceutics. 2007;**4**(6):826-832

[63] Nielsen IL, Chee WS, Poulsen L, Offord-Cavin E, Rasmussen SE, Frederiksen H, et al. Bioavailability is improved by enzymatic modification of the citrus flavonoid hesperidin in humans: A randomized, double-blind, crossover trial. The Journal of Nutrition. 2006;**136**(2):404-408

[64] Blanco E, Shen H, Ferrari M. Principles of nanoparticle design for overcoming biological barriers to drug delivery. Nature Biotechnology. 2015;**33**(9):941-951

[65] Kou L, Bhutia YD, Yao Q, He Z, Sun J, Ganapathy V. Transporter-guided delivery of nanoparticles to improve drug permeation across cellular barriers and drug exposure to selective cell types. Frontiers in Pharmacology. 2018;**9**:27

[66] Wang L, Fu R, Meng Y, Liang J, Xue W, Hu H, et al. pH sensitive quercetin nanoparticles ameliorate DSS-induced colitis in mice by colon-specific delivery. Molecular Nutrition & Food Research. 2023;**11**(7):e2300051

[67] Jing S, Chen H, Liu E, Zhang M, Zeng F, Shen H, et al. Oral pectin/oligochitosan microspheres for colon-specific controlled release of quercetin to treat inflammatory bowel disease. Carbohydrate Polymers. 2023;**316**:121025

Chapter 8

Preclinical Therapeutic Effects of Quercetin on Gastrointestinal Cancers

Bernadette Xin Jie Tune, Yuan Seng Wu, Rhanye Mac Guad, Aimi Syamima Abd Manap, Sheryar Afzal, Kalaivani Batumalaie, Ker Woon Choy, Neeraj Kumar Fuloria, Shivkanya Fuloria, Vetriselvan Subramaniyan and Mahendran Sekar

Abstract

Gastrointestinal (GI) cancers were included in the top 10 most common cancers in 2020. Future incidences are expected to rise due to the varying risk factors and aetiologies. With high incidences and mortality rates, current cancer treatments fail to reduce mortality and morbidity in GI cancer patients. A large part of cancer research has been geared towards targeted and personalized medicine, although effective, it may not be the most cost-friendly and feasible option to treat patients from varying socioeconomic backgrounds. Hence, natural compounds may present as an attractive alternative treatment in the management of GI cancers. Quercetin is a well-known flavonoid compound, found in almost all fruits and vegetables. It has also been widely studied for its anticancer properties, such as anti-oxidative, anti-inflammatory, anti-proliferative and anti-angiogenic properties. In this chapter, the authors discuss the potential of quercetin in treating GI cancers, which includes the biosafety and toxicity of quercetin, applications of quercetin in common GI cancers, such as gastric, hepatic, colorectal, pancreatic and oesophageal cancers, along with the corresponding molecular mechanisms. The authors also present evidences of quercetin as an adjuvant therapeutic agent with other anticancer drugs.

Keywords: gastrointestinal cancers, quercetin, natural products, cancer treatment, molecular mechanism

1. Introduction

Gastrointestinal (GI) cancers constitute a formidable health challenge globally, exhibit varying prevalence, epidemiological patterns and mortality rates. Colorectal cancer, with an estimated 1.9 million new cases in 2020, ranks among the most prevalent type of cancer, particularly in developed nations, which is largely influenced by lifestyle factors [1]. Gastric cancer accounts for 1.1 million new cases annually and displays diverse epidemiology linked to regional variations and risk factors such as *Helicobacter pylori* infection and poor dietary practices [2]. Liver cancer, which was responsible for over 830,000 new cases, is notably prevalent in East Asia and sub-Saharan Africa, primarily associated with chronic hepatitis B and C infections [3]. Oesophageal cancer, with approximately 604,000 new cases reported in 2020, exhibits distinct geographic disparities, emphasizing the impact of local risk factors [3]. High mortality rates characterize GI cancers collectively, with liver cancer ranking as the third leading cause of cancer-related deaths globally [3]. These statistics highlight the urgency of evidence-based interventions and targeted research initiatives to address the complex interplay of factors contributing to the prevalence and mortality rates of GI cancers.

The journey towards identifying novel therapeutic strategies for GI cancers begins with understanding current treatments' shortcomings. Despite advancements in medical science, such as surgical interventions, chemotherapy and targeted therapies, current treatment modalities have not yielded a substantial reduction in mortality and morbidity associated with these malignancies [4]. One glaring example is colorectal cancer, a prevalent GI malignancy that continues to exert a significant toll on global health. Despite extensive research and therapeutic interventions, colorectal cancer remains a leading cause of cancer-related deaths worldwide [5].

The limitations of existing treatments highlight the pressing need for innovative and more effective therapeutic strategies. Responding to this imperative, attention has turned towards natural compounds with potential therapeutic benefits [6]. One compound that has emerged in the spotlight is quercetin, a polyphenolic flavonoid abundantly found in fruits, vegetables and certain beverages [7]. Quercetin has garnered attention not only for its antioxidant and anti-inflammatory properties but also for its intriguing anticancer potential, making it a subject of intense scientific investigation [8]. Extensive preclinical research has explored the potential of quercetin in inhibiting the growth of cancer cells, inducing apoptosis and suppressing the development of new blood vessels that support tumour growth [9]. Importantly, quercetin has been reported to enhance the efficacy of conventional chemotherapy agents when combined, potentially overcoming resistance mechanisms [10].

The appeal of quercetin lies not only in its anticancer properties but also in its favourable safety profile. Unlike some conventional cancer treatments associated with significant toxicities, quercetin has demonstrated low toxicity in various preclinical and clinical studies [11]. This characteristic is particularly crucial in cancer treatment where minimizing adverse effects on normal tissues is paramount for improving patients' quality of life during and after therapy [11]. In addition to its direct effects on cancer cells, quercetin has been investigated for its potential to modulate the tumour microenvironment, which plays a critical role in cancer progression. Interventions that can modify this milieu hold promise for disrupting the supportive conditions for tumour growth and metastasis [12]. Quercetin's ability to influence the tumour microenvironment further expands its potential therapeutic impact. Its molecular structure, depicted in **Figure 1**, lays the foundation for understanding its multifaceted interactions within biological systems. Its chemical composition contributes to its ability to modulate cellular pathways, making it an intriguing candidate for cancer therapy [13].

This chapter aims to comprehensively review and analyse the preclinical therapeutic effects of quercetin against GI cancers, shedding light on its potential as a therapeutic agent. We seek to provide a thorough overview of quercetin's biosafety

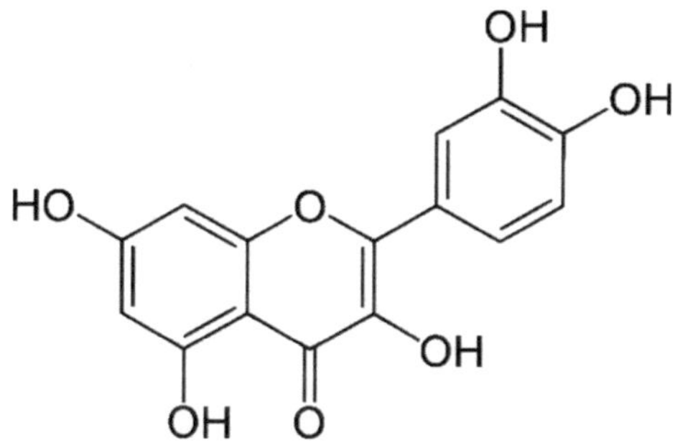

Figure 1.
Chemical structure of quercetin.

and toxicity profiles, ensuring a nuanced understanding of its safety considerations, essential for assessing its potential as a therapeutic agent. Then, exploring quercetin's pharmacokinetics will elucidate its journey from laboratory studies to potential clinical applications. Understanding the pharmacokinetics of quercetin is crucial, and this section will encompass *in vitro* and *in vivo* studies, providing a holistic perspective on its absorption, distribution, metabolism and excretion. Finally, the core of the chapter details the preclinical therapeutic effects and molecular mechanisms underlying quercetin's efficacy against GI cancers, namely gastric, liver, colorectal, pancreatic and oesophageal cancers. Each cancer type is comprehensively examined in *in vitro* and *in vivo* studies, emphasizing key findings related to cell viability, proliferation, migration/invasion, chemoresistance, angiogenesis and other relevant parameters. Through this detailed analysis, it aims to provide a comprehensive resource for researchers, clinicians and policymakers interested in the potential therapeutic role of quercetin in the management of GI cancers.

Pursuing novel and effective therapeutic strategies for GI cancers is imperative, given the persistent challenges associated with current treatments. With its multifaceted properties and promising preclinical data, quercetin represents a beacon of hope in the quest for innovative cancer therapies. As research advances and clinical trials progress, the true potential of quercetin in reshaping the landscape of GI cancer treatment may unfold, offering new avenues for improving patient outcomes and addressing the unmet needs in cancer care.

2. Biosafety and toxicity profiles of quercetin

Quercetin possesses a variety of pharmacological activities, and for further medical application, it is important to evaluate its biosafety and toxicity profiles in experimental and clinical studies. With oral administration to mice and rats, quercetin consistently did not induce any significant changes in several mutagenicity/genotoxicity endpoints, such as micronuclei, chromosomal aberrations, sister chromatid exchange, unscheduled DNA synthesis and alkali-labile DNA damage in somatic cells [14]. Furthermore, based on the Ames test and combined *in vivo* micronucleus and comet assay, there was no induction of *in vivo* genotoxic potential or indication of any oxidative DNA damage, suggesting the absence of mutagenicity and genotoxicity potential of the isoquercitrin-γ-cyclodextrin (IQC-γCD) inclusion complex in rat liver tissues [15].

On the other hand, a clinical study [16] indicated that total sperm motility was significantly inhibited following exposure to 100, 200 and 400 μM quercetin for 6 and 12 hours in a dose-dependent manner, as compared to the controls ($p < 0.05$), suggesting that quercetin inhibits sperm functions. Another clinical study showed that extra quercetin consumption was able to lower blood pressure in adult hypertensive patients after receiving quercetin at dosages ranging from 162 to 1095 mg/day for a period of 7 days to 6 weeks [17]. A study also indicated that, after 1 week of treatment, 16 patients of the quercetin-treated group tested negative for severe acute respiratory syndrome coronavirus 2 (SARS-CoV-2), of which 12 patients had all their symptoms diminished; in the control group, two patients tested negative and four patients had their symptoms partially improved. These findings suggest that quercetin significantly shortens the timing of molecular test conversion from positive to negative ($p < 0.05$), at the same time reducing symptom severity and negative predictors of COVID-19 [18]. Moreover, it was observed that disease biomarkers, including lactate dehydrogenase (LDH) (−35.5%), ferritin (−40%), C-reactive protein (CRP) (−54.8%) and D-dimer (−11.9%), were reduced, indicating SARS-CoV-2 viral clearance. According to a study on patients with chronic obstructive pulmonary disease (COPD) who had mild-to-severe lung disease and a first second of forced expiration (FVE1) ranging between >35% and <80%, no patients experienced any drug-related severe adverse events after supplemented with quercetin at 500, 1000 or 2000 mg/day in a dose-escalation manner for 1 week [19], suggesting that quercetin was safely tolerated up to 2000 mg/day, as assessed by lung function and blood profile. These studies provided sufficient data on quercetin biosafety and tolerance in patients while effectively administering therapeutic effects.

3. Pharmacokinetics of quercetin

Quercetin exhibits established pharmacokinetic characteristics in both laboratory animals and humans. The previous research finding suggested that unmodified quercetin material has limited bioavailability when administered orally, restricting its practical application in clinical settings. For example, quercetin is consumed in the glycoside form, with glycosyl groups being liberated during chewing, digestion and absorption. Subsequently, the enzymes known as β-glycosidases transform quercetin glycosides into aglycone in the intestines, prior to their absorption into enterocytes [20, 21]. Prior research has also indicated the involvement of oral and intestinal microorganisms in this enzymatic hydrolysis. Since quercetin is a lipophilic substance, it is thought to be able to pass through intestinal membranes by simple diffusion. In general, this absorption is preferable to those of its glycoside counterparts that enter the intestines unaltered [20, 21].

As of now, the bioavailability of quercetin glycosides has been investigated in several human studies. For example, a previous clinical study reported pharmacokinetic characteristics of corvitin (quercetin) in healthy participants. The study shows that the primary metabolites of quercetin are its methoxy derivatives, specifically isorhamnetin and tamarixetin, together with their conjugates of glucurone and sulfatide. About 20% of quercetin dosages are subject to metabolic changes in 20 minutes, free quercetin conjugates by 33% and isorhamnetin by 70% in 20–30 minutes. Corvitin is eliminated with urine as conjugates of quercetin and isorhamnetin, with an average maximal clearance rate of 1.39 mg/h occurring 1.63 hours after drug administration [22]. Hollman et al. reported that patients with ileostomy exhibited a higher absorption percentage of quercetin glycosides from onions compared to the pure aglycone form [23].

On the other hand, Mansour et al. reported that individuals who had eaten an onion-containing lunch had ileostomy fluid that included considerable levels of aglycone. A significant amount of quercetin aglycone and a small amount of quercetin glycosides were found in the fluid. One conceivable reason is that β-glycosidase enzymes hydrolyse quercetin glycosides, converting them to aglycone. Most of these enzymes are found in the GI tract, where the enzymes are then absorbed [24]. Another study looked at the excretion of quercetin metabolites in young people after eating gently fried onions. It was determined that the fraction of quercetin intermediates discharged in urine following dietary quercetin consumption is approximately 4.7% of the whole intake, which is 12.9 mol. The makeup of metabolites excreted in urine differed significantly from that of plasma. The majority of the quercetin intermediates in the urine were quercetin glucuronide sulphates, quercetin diglucuronides, with additional significant metabolites including isorhamnetin-3-glucuronide and quercetin 3′-glucuronide [25].

Moreover, a study reported the pharmacokinetic characteristics of intravenous administration of quercetin in cancer patients at dosages ranging from 60 to 2000 mg/m^2. The researchers determined that the safe amount of quercetin was 945 mg/m^2, but the hazardous level was reported to produce nephrotoxicity, hypertension, vomiting and a drop in blood potassium. Intravenous quercetin was discovered to have the following half-lives: 3.8–86 minutes for its elimination half-life and 0.7–7.8 minutes for its distribution; 0.23–0.84 L/min/m^2 for its clearance and 3.7 L/m^2 for its distribution volume [26]. Following that, experimental research conducted on rabbits showed that the oral and parenteral treatment of quercetin along with pectin enhanced GI absorption by about 11 times and bioavailability by about 10 times compared to quercetin alone. According to the study, quercetin's pharmacokinetic characteristics and solubility are improved when combined with pectin for medicinal purposes [20]. Subsequent research conducted by Salehi et al. revealed quercetin's pharmacokinetic properties, indicating that its T_{max} and C_{max} were 0.7 ± 0.3 hour and 2.3 ± 1.5 µg/ml, respectively, at a dose of up to 200 mg [8].

4. Preclinical therapeutic effects and associated molecular mechanisms of quercetin against gastrointestinal cancers

Gastrointestinal (GI) cancers refer to cancers occurring in the GI tract, including liver and pancreatic cancers. The most common types of GI cancers are colorectal, gastric, liver, oesophageal and pancreatic cancers [27]. According to a report in 2020, GI cancers were among the top 10 most common cancers occurring worldwide. Colorectal cancer ranks 3rd in occurrences and 2nd in mortality, gastric and liver cancers ranked 5th in incidences, while oesophageal cancers ranked 7th in incidences [3].

Gastric cancers arise from malignancies of the stomach; there are two classifications of gastric cancers depending on the topographical subsite of the tumour, which are cardia and non-cardia gastric cancers, referring to the upper stomach and the lower stomach, respectively [3, 27]. Both of these subtypes present with different risk factors, carcinogenesis and epidemiology. For example, non-cardia gastric cancers are primarily attributed to *H. pylori* infections [28, 29]. Cardia gastric cancers, on the other hand, have similar risk factors as that of oesophageal adenocarcinoma (gastroesophageal reflux disease and obesity) [27].

Liver cancers were reported to be more prevalent in men than in women, with rates of both incidences and mortality being two and three times higher in men than in women [3]. Several more common types of liver cancers include hepatocellular carcinoma (HCC) and intrahepatic cholangiocarcinoma (ICC). Common risk factors of HCC include Hepatitis B virus (HBV) and Hepatitis C virus (HCV) infections, aflatoxins, heavy alcohol intake, obesity, type 2 diabetes and use of tobacco [3, 27]. Meanwhile, incidences of ICC have been linked to primary sclerosing cholangitis (PSC), liver fluke infections, fibropolycystic liver disease and biliary stones [27, 30].

Colorectal cancer was observed to be one of the most common cancers in incidences, ranking 3rd overall, with 2nd in terms of mortality. One observed trend in colorectal cancer is the shift in socioeconomic development, with incidence rates higher in transitioned countries than transitioning countries [3]. Changes in lifestyle, such as increased meat consumption, reduced physical activity and excess body weight, may be among the risk factors for colorectal cancer [31]. Other risk factors include increased alcohol intake, heavy tobacco use and increased consumption of processed or red meats [3].

Pancreatic cancer was reported to be the 7th leading cause of cancer death in males and females, largely due to its poor prognosis. Incidence rates were up to 4- to 5-fold higher in developed countries, with the highest incidences recorded in Europe, North America and Australia/New Zealand [3]. Risk factors of pancreatic cancer include obesity, diabetes and heavy alcohol consumption [27].

Oesophageal cancer ranks 7th in terms of incidence and 6th in mortality worldwide. The most common subtypes of oesophageal cancer are oesophageal squamous cell carcinoma (ESCC) and oesophageal adenocarcinoma (EAC). The geographical distribution of the subtypes may signify the corresponding risk factors associated with the type of cancer. ESCC was more prevalent in Eastern Asia (largely in China), Southern Africa and Eastern Africa. Possible risk factors may include nutritional deficiencies and exposure to nitrosamines [32]. However, other risk factors include heavy tobacco and alcohol use. While EAC was reported to be more prevalent in Western countries. Incidences in high-income countries were attributed to excess body weight and incidences of gastroesophageal reflux disorder [3].

With the common occurrence of GI cancers, developing new strategies to combat and manage GI cancers can be advantageous. Using natural products in medicine has also been a well-known strategy in drug discovery. Examples of common natural products in cancer treatment are taxanes, vinca alkaloids and anthracyclines [33]. Quercetin is a flavonoid found in most fruits and vegetables, possibly one of the most common dietary flavonols [34]. Quercetin can exert antioxidant effects at low doses, which provide a chemopreventive effect, yet also exert a pro-oxidant effect at higher doses, providing a chemotherapeutic effect [35].

4.1 Gastric cancer

Quercetin's chemopreventive effects, demonstrated in both *in vitro* and *in vivo* models, make it a flavonoid with great potential in oncology. The effects of quercetin are biphasic and dose-dependent. Quercetin exhibits chemopreventive effects when present in low concentrations, owing to its antioxidant properties. However, when found in high concentrations, quercetin serves as a pro-oxidant and might potentially induce chemotherapeutic effects [35]. Quercetin's anticancer properties stem from its capacity to modify cyclins, pro-apoptotic, phosphatidylinositol 3-kinase (PI3K)/protein kinase B (Akt) and mitogen-activated protein kinase (MAPK) molecular

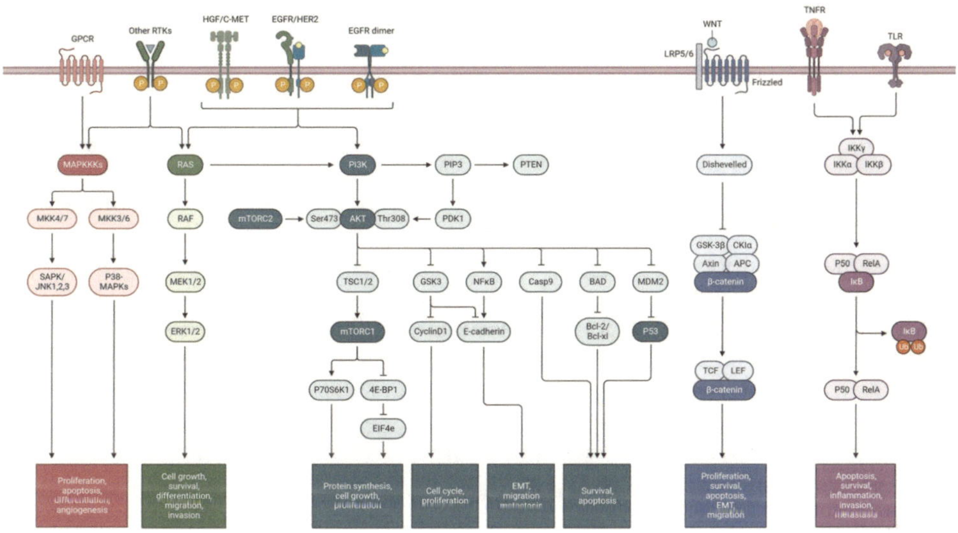

Figure 2.
Signalling pathway in gastric cancer.

pathways, thereby reducing proliferation, inducing apoptosis, causing cell cycle arrest and inhibiting mitotic processes. **Figure 2** shows signalling pathways in gastric cancer, which are commonly targeted by therapeutic intervention. Quercetin's anticancer effects on gastric cancer have been the subject of numerous studies, which will be covered in detail in this section.

Zeng et al. conducted a bioinformatics analysis to assess the impact of quercetin on gastric cancer. The study aimed to discover the specific targets of quercetin in NCI-N87 gastric cancer cells exposed to this compound. The researchers exposed human gastric cancer cells (NCI-N87) to a concentration of 15 μM quercetin for 48 hours, using dimethyl sulphoxide (DMSO) as a control. The HiSeq 2500 DNA sequence data were utilized to compare differentially expressed genes (DEGs) between groups. A sophisticated technique was employed to evaluate the protein-protein interaction (PPI) network. The regulatory network of transcription factors (TFs)-DEGs was determined using Cytoscape. The DEGs identified in the study were Fos proto-oncogene (FOS, degree = 12), aryl hydrocarbon receptor (AHR, degree = 12), Jun proto-oncogene (JUN, degree = 11) and cytochrome P450 family 1 subfamily A member 1 (CYP1A1, degree = 11). These DEGs showed significant associations with other proteins in the protein-protein interaction (PPI) network, particularly with proteins having higher degrees. The expression levels of Early growth response 1 (EGR1), FOS like 1 (FOSL1), FOS and JUN were elevated among the five TFs-DEGs, while AHR was downregulated. The Wnt signalling pathway exhibited an abundance of FOSL1, JUN and Wnt family member 7B (WNT7B). CYP1A1 exhibited a strong association with AHR in the PPI network. Thus, quercetin may have specifically affected FOS, AHR, JUN, CYP1A1, EGR1, FOSL1 and WNT7B in gastric cancer [36].

Apart from that, quercetin has also been investigated in combination studies. In a study conducted by Lei et al., the effectiveness of a combination of quercetin and irinotecan in mitigating the spread of gastric cancer was examined. This investigation was done by assessing the expression of genes and proteins [37]. This study compared the effects on β-catenin expression, cell viability and apoptosis of low-dose SN-38 (irinotecan metabolite) in combination with quercetin and high-dose SN-38 alone. In addition, the effects of quercetin and low-dose irinotecan on gastric cancer metastasis were examined using *in vivo* xenograft animal models and *in vitro* investigations. The results showed that the human gastric adenocarcinoma (AGS) cells treated with a combination of quercetin and a low dose of SN-38 exhibited reduced levels of the β-catenin protein compared to the treatment with quercetin alone. The expression of integrin β6 (ITG-β6) and Twist-1 genes which are two markers of epithelial-mesenchymal transition (EMT), as well as the expression of cyclooxygenase-2 (COX-2) gene, was shown to be higher in cells treated with a high dose of irinotecan compared to cells treated with combination therapy. Following combination therapy, the AGS mouse model exhibited a notable decrease in VEGF-A (vascular endothelial growth factor A), VEGF-receptor 2 (VEGFR-2) and the proportion of Tek tyrosine-protein kinase receptor (Tie2)-expressing monocytes (TEMs). According to the results, the effectiveness of irinotecan in treating gastric cancer (GC) could be enhanced by combining it with quercetin [37].

A distinct investigation was conducted to assess the mechanism and impact of quercetin on the spread of gastric cancer. Additionally, the study aimed to determine if the involvement of urokinase plasminogen activator and urokinase plasminogen receptor (uPA/uPAR) played a role in this process [38]. The purpose of the study was to determine whether quercetin could have an impact on the uPA/uPAR system, which is crucial to gastric cancer metastasis. Precancerous tissues' uPA and uPAR activity levels were measured, and the results were compared to the migration and invasion patterns of various gastric cancer cell lines [37]. According to the data, uPA and uPAR activities in precancerous tissues were lower than those in gastric cancer cells, and there was a correlation between uPAR expression and gastric cancer cell line migration and invasion. After gastric cancer BGC823 and AGS cells were treated with quercetin (10 μM for 72 hours), uPA and uPAR protein expression levels were decreased along with migration and invasion. The combination of quercetin and uPAR knockdown reduced in matrix metalloproteinase-9 (MMP-9) and matrix metalloproteinase-2 (MMP-2) activities, effectively inhibiting p21-activated kinase 1 (Pak1)-LIM kinase 1 (Limk1)-cofilin signalling. Treatment with quercetin suppressed the activation of adenosine monophosphate (AMP)-activated protein kinase α (AMPKα), nuclear factor-kappa B (NF-κB), extracellular signal-regulated kinase 1/2 (ERK1/2) and protein kinase C-delta (PKC-δ), decreasing uPA and uPAR expression. Ultimately, quercetin can potentially turn out to be an emerging factor in the treatment of gastric cancer, effectively diminishing metastasis and invasion [38].

Furthermore, *in vitro* experiments were conducted to investigate the ability of quercetin to promote cell death, specifically in human gastric cancer cells. These experiments also examined the impact of quercetin on apoptosis and gene expression. Analysis of their results demonstrated that quercetin can trigger apoptosis in GC cells and modulate gene expression [39]. According to flow cytometry, quercetin exacerbated AGS cells to undergo apoptosis, reduced certain protein levels that prevented the mitochondrial membrane from being intact and elevated reactive oxygen species (ROS). By using Western blotting, it was possible to observe that quercetin increased the levels of pro-apoptotic proteins like B-cell lymphoma protein 2 (Bcl-2)-associated X protein (Bax), Bcl-2-associated death promoter (Bad) and Bcl-2-interacting domain death agonist (Bid) while decreasing the levels of anti-apoptotic proteins like Bcl-x (B-cell lymphoma x protein), B-cell lymphoma 2 (Bcl-2) and myeloid cell leukaemia-1 (Mcl-1). Moreover, quercetin elicited diverse impacts on gene expression.

For example, quercetin reduced the levels of KDELC2F (KDEL [Lys-Asp-Glu-Leu] containing 2), vascular endothelial growth factor B (VEGF-B) and cyclin-dependent kinase 10 (CDK10) while increasing the levels of tumour protein p53-induced nuclear protein 1 (TP53INP1), tumour necrosis factor receptor superfamily member 10D (TNFRSF10D) and JUN-B. Eventually, their research demonstrated the molecular process, gene expression and signalling pathway entailed in quercetin's capacity to cause human gastric cancer cells to undergo apoptosis [39].

The impact of anti-proliferative drugs (doxorubicin (DOX) and quercetin) and small interfering RNA (si-RNA) directed against CDC20 (cell division cycle protein 20 homolog) on GC was assessed by Hemati et al. The researchers examined the use of niosome-encapsulated delivery vehicles to deliver si-RNA and medicines. It was discovered that the administration of si-RNA, together with anticancer medications, resulted in the suppression of CDC20 expression, hence enhancing the effectiveness of gastric cancer treatment [40]. In order to enhance the loading capacity and improve the physicochemical properties of si-RNA, the researchers altered the amount of cationic lipid in cationic PEGylated niosomes. Quercetin, DOX and anti-CDC20 si-RNA were incorporated into the co-delivery system. The system was then evaluated for its physicochemical qualities, controlled release, thermosensitivity, rates of apoptosis and gene silencing efficacy. Interestingly, the data showed that the co-delivery system, intended explicitly for loading si-RNA, had a suitable and significant positive charge for delivering drugs. In addition, the researchers demonstrated a thermosensitive drug release pattern that effectively suppressed CDC20 expression,

Gastrointestinal cancer	Preclinical model (*in vitro*/*in vivo*)	Anticancer mode	Molecular mechanism	Reference
Gastric cancer	*In vitro* (NCI-N87 cells)	Anti-proliferative effects	EGR1, FOS-1, FOS and JUN expression levels were elevated while AHR was downregulated.	[36]
	In vitro (AGS cells) and *in vivo*	Inhibition on cell viability, anti-apoptosis and anti-proliferation of gastric cancer metastasis	The combination of quercetin and a low dose of SN-38 reduced β-catenin protein levels in AGS cells compared to quercetin alone.	[37]
	In vitro (BGC823 AGS cells)	Anti-proliferation of gastric cancer metastasis	Inhibition of uPA, uPAR and downstream target expression	[38]
	In vitro (AGS cells)	Apoptosis induction	Apoptosis induction via ROS increased the levels of pro-apoptotic proteins (Bax, Bad and Bid), decreasing the levels of anti-apoptotic proteins (Bcl-x, Bcl-2 and Mcl-1)	[39]
	In vitro (AGS cells)	Inhibition on gastric cancer cell growth	Suppression of CDC20 by si-RNA which inhibited cancer cell growth	[40]

Table 1.
Summary of the latest study on the therapeutic effect of quercetin in gastric cancer.

surpassing the individual delivery of either si-RNA or the drug alone. Additionally, their approach efficiently suppressed the proliferation of gastric cancer cells. Their findings indicated that PEGylated niosomes containing both CDC20 si-RNA and anticancer medicines could serve as an innovative approach for treating gastric cancer [40]. A few studies on quercetin's therapeutic effects in gastric cancer are listed in **Table 1**.

4.2 Liver cancer

Fruits, vegetables and plants all contain quercetin having anti-inflammatory, anti-cancer and antioxidant qualities. Its advantageous properties, including hepatoprotective benefits against liver diseases, have been established in various human illnesses. Hepatocellular carcinoma (HCC) is a primary liver tumour with a high death rate and a delayed diagnosis, which makes it a promising target for research on quercetin effects. HCC is the most prevalent type of primary liver cancer and the sixth most often occurring tumour globally, making it the fourth most deadly neoplasm [41]. Hepatitis C and B virus (HCV and HBV, respectively) and other aetiological agents cause liver damage that leads to the development of HCC through the phases of liver fibrosis and cirrhosis, which can take years or even decades to manifest. Due to its intricate aetiology and molecular heterogeneity, curative therapy for liver cancer is met with challenges [41, 42]. In the first-line scenario for advanced liver cancer, systemic treatment is employed in these instances, employing the two tyrosine kinase inhibitors (TKIs) that are currently available: lenvatinib and sorafenib [43]. After prolonged treatment, liver cancer cells can become resistant to sorafenib, regardless of its efficacy [42], where several TKIs (regorafenib and cabozantinib) and monoclonal antibody have been approved by the FDA for HCC previously treated with sorafenib. Several studies have examined the anticancer benefits of natural substances against liver cancers, such as resveratrol, curcumin and melatonin, to reduce the toxicity and adverse responses brought on by these chemotherapeutic drugs [44].

According to Yamada et al. [45], myricetin and quercetin can block the protein kinase B (AKT) signalling axis, which may subsequently impede HuH7 cell migration driven by transforming growth factor-α (TGF-α) and hepatocyte growth factor (HGF). The study examined how quercetin affected HuH7 cancer cells' migration in response to TGF-α or HGF. In a dose-dependent manner, quercetin significantly impeded HuH7 cell migration induced by both TGF-α and HGF. Moreover, myricetin, an additional flavonol molecule, significantly inhibited the migration of cancer cells. TGF-α- and HGF-mediated receptor autophosphorylation was unaffected by myricetin or quercetin. Furthermore, quercetin did not affect the TGF-α- or HGF-induced activation of p38 MAPK. However, the growth factors-mediated AKT phosphorylation was inhibited by myricetin and quercetin. Through inhibiting the AKT signalling pathway but not p38 MAPK, their study demonstrated that quercetin might reduce growth factor-driven cell migration of liver cancer [45].

Quercetin can reduce phosphorylation of ERK1/2 and limit proteasome activity, as demonstrated by Ding et al. Proteasome chymotrypsin-like activity rose in response to increasing ERK1/2 activity, while proteasome chymotrypsin-like activity decreased in response to increased MEK1 (mitogen-activated protein kinase kinase 1) activity. The administration of quercetin reduced the amount of proteasome β subunits expressed. According to Ding et al., their investigation demonstrated that the proteasome β subunits' expression may be modulated by the MEK1/ERK1/2 signalling pathway, resulting in a decrease in the proteasome's chymotrypsin-like activity [46]. HepG2 cells continued to exhibit steady levels of caspase and trypsin-like protease activity as well as elevated c-Jun N-terminal kinase (JNK) and p38 MAPK activity and reduced phosphorylation of ERK1/2. The inhibition of the JNK and p38 MAPK signalling pathways may not repair the reduced proteasome activity caused by quercetin. Proteasome chymotrypsin-like activity increased or decreased depending on whether MEK1 was upregulated or downregulated. MEK1/ERK1/2 inhibition and quercetin treatment decreased the expression of the proteasome's β subunits [46].

Quercetin targets apoptosis by upregulating Bax, caspase-3 and p21 and downregulating Akt, polo-like kinase 1 (PLK-1), cyclin-B1, cyclin-A, cell division control 2 (CDC-2), cyclin-dependent kinase 2 (CDK-2) and Bcl-2. It has been observed to decrease signal transducer and activator of transcription 3 (STAT-3) activation and promote STAT-3 protein degradation in liver cancer cells. Numerous studies have demonstrated quercetin's anticancer properties, suggesting a possible role for the compound in chemoprevention. When quercetin was combined with other anticancer substances, it demonstrated synergistic benefits. Research has demonstrated that quercetin and 5-fluorouracil (5-FU) work synergistically in liver cancer cell lines. Compared to quercetin treatment alone, this combination resulted in increased growth inhibition in some cell lines [47]. Another research found that the anti-inflammatory, proliferative and angiogenesis-related genes TNF-α, VEGF, P53 and NF-κB were all downregulated by quercetin, either alone or in combination with sorafenib, the first medication licensed to treat advanced liver cancer. Treatment with sorafenib plus quercetin resulted in cell cycle arrest, apoptosis and necrosis, in addition to a considerable inhibition of liver tumour growth [48]. Research on the synergistic antitumour effects of quercetin and oncolytic adenovirus in liver cancer cells suggests that quercetin might enhance oncolytic adenovirus ZD55-TRAIL (tumour necrosis factor-related apoptosis-inducing ligand)-mediated growth inhibition and death in liver cancer cells. *In vitro* and *in vivo* antiliver cancer experiments have demonstrated the potential of quercetin combination therapy [49].

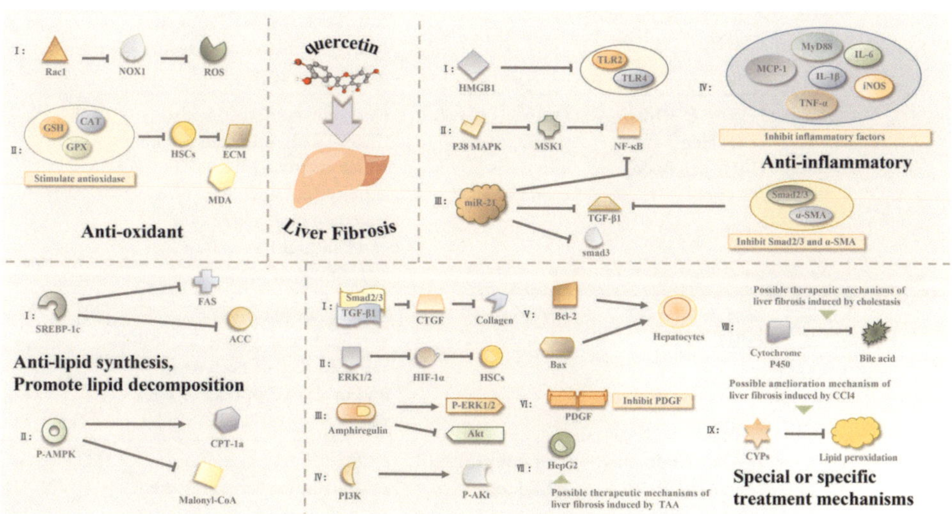

Figure 3.
Diagram of the mechanism of quercetin in the treatment of liver fibrosis. Source: Adapted from Ref. [50].

These findings point to a distinct anti-proliferative and pro-apoptotic action of quercetin in liver cancer as well as a possible regulating function for the tumour cell cycle progression that requires more research. Through the modification of several cellular processes and signalling pathways, this flavonoid appears to have antitumoural efficaciousness; nevertheless, more research is necessary to fully understand its mechanisms of action against liver cancer. In addition to the emerging usage of quercetin derivatives with anticancer efficaciousness, emerging tactics of combination and drug-delivery systems may enhance such features of cancer inhibition and expand treatment choices for patients with liver cancer [44]. **Figure 3** illustrates the

Gastrointestinal cancer	Preclinical model (*in vitro*/*in vivo*)	Anticancer mode	Molecular mechanism	Reference
Liver cancer	*In vitro* (SMMC-7721, BEL-7402 liver cancer cells, LO2 normal liver cells)	Inhibition of cell proliferation	Inhibition of Akt/mTOR, reduced glucose uptake and lactate production, reduced p-Akt/Akt and p-mTOR/mTOR rates, increased cell growth inhibition	[51]
	In vitro (SMMC-7721 and HepG2 liver cancer cells LO2 normal liver cells)	Cell growth inhibition, autophagy induction, apoptosis induction	Akt/mTOR inhibition and MAPK activation lead to increase in LC3A/LC3B-II and Beclin1 protein levels and decrease in p62 protein expression, increase in the percentage of apoptotic cells	[52]
	In vitro (BEL-7402 cells, multidrug-resistant cell line BEL/5-FU)	Chemosensitizing effect towards 5-FU	FZD7/β-catenin inhibition, reduction in ABCB1, ABCC1 and ABCC2 expression, as well as inhibition of ABCC2 function	[53]
	In vitro (HepG2 cells)	Apoptosis induction	MEK1/ERK1/2 inhibition, increased levels of cleaved caspase-3, cleaved PARP and Bax proteins, reduced levels of Bcl-2 protein	[46]
	In vitro (SMMC-7721, HepG2 and HuH7 cell lines)	Decrease of cell proliferation, apoptosis induction	Quercetin combined with ZD55-TRAIL inhibited NF-κB which increased apoptotic activity (nuclear fragmentation, chromatin condensation) as well as apoptotic markers	[49]
	In vitro (HepG2 cells)	Reduced cell viability	Synergistic activity of sorafenib and quercetin, with reduced IC_{50} of both drugs and increased uptake of both drugs	[54]
	In vivo (Nude mice subcutaneously injected with LM3 cells)	Tumour growth inhibition	Quercetin 100 mg/kg was orally administered, reduced in tumour volume, increased necrosis and TUNEL-positive cells	[51]

Gastrointestinal cancer	Preclinical model (*in vitro*/*in vivo*)	Anticancer mode	Molecular mechanism	Reference
		In vivo (SMMC-7721 xenograft mouse model)	Quercetin 50 mg/kg administered via intraperitoneal injection, reduction in tumour size, HK2 and Ki67 protein expression and p-Akt and p-mTOR protein levels observed	[51]
	In vivo (Nude mice subcutaneously injected with SMMC-7221 cells)	Suppression of tumour growth, apoptosis and autophagy induction	Quercetin 60 mg/kg administered orally, reduction in tumour weight and volume observed, as well as apoptotic markers (increased Bax, cleaved caspase-3, reduced Bcl-2 protein expression)	[52]
	In vivo (MHCC97H xenograft mouse model)	Suppression of tumour growth and progression, apoptosis induction	Quercetin encapsulation (PLGA-loaded gold-quercetin nanoparticles) 30, 40 and 50 mg/kg via intraperitoneal injection causes inhibition of AP-2β/hTERT, p50/NF-κB/COX-2, Akt/ERK1/2 pathways, reduced tumour weight and volume and increased apoptotic activity observed	[55]
	In vivo (Mice subcutaneously injected with HuH7 cells)	Inhibition of tumour growth	Quercetin combined with dasatinib 50 mg/kg of quercetin with 5 mg/kg of dasatinib administered orally halted tumour progression	[56]

Table 2
Basic characteristics of in vitro *and* in vivo *studies using quercetin in liver cancer.*

common mechanistic pathways of quercetin. A few research on quercetin's therapeutic benefits in liver cancer are listed in **Table 2**.

4.3 Colorectal cancer

Quercetin has garnered significant interest in its ability to prevent and combat colorectal cancer [57]. Numerous research have examined these effects both in laboratory settings (*in vitro*) [57–60] and in living organisms (*in vivo*) [61–64]. This section will provide a concise overview of the anticancer effects of quercetin, followed by a summary of its potential as a novel therapeutic agent for treating colorectal cancer.

Evaluations have been conducted on the role of quercetin in apoptosis, hyperproliferation and inflammation, as well as its mechanism in 1,2-dimethylhydrazine (DMH)-induced carcinogenicity and tumour multiplicity [65]. Rats were orally given quercetin at doses of 25 or 50 mg/kg body weight and subcutaneously injected with

20 mg/kg body weight of DMH for 15 weeks. Afterwards, the rats were euthanized. The DMH generates reactive oxygen species (specifically superoxide) through the process of membrane lipid peroxidation, leading to a disruption in redox equilibrium. DMH also reduces the amount of antioxidants in tissues. The DMH-induced intestinal carcinogenicity led to an increase in proliferative and inflammatory variables due to a low Bax/Bcl-2 ratio and dysregulation of apoptosis. Pretreatment with quercetin mitigated the deleterious effects of DMH, such as maintaining the activity of detoxifying enzymes and decreasing proliferation as well as early indicators (such as mucin depletion and goblet cell disintegration) in the colonic tissue. Quercetin controlled the expression of β-catenin and adenomatous polyposis coli (APC), decreasing in the occurrence and number of tumours. The histology findings confirm the beneficial effect of quercetin in mitigating the pathogenic changes generated by DMH [65].

A prior investigation similarly assessed the impact of daily intake of microencapsulated probiotics either alone or in conjunction with microencapsulated quercetin to prevent colorectal cancer. The researchers utilized ApcMin/+ mice which exhibit spontaneous intestinal adenomas and cancer. The researchers evaluated changes in tissue structure, intestinal bleeding, fat accumulation, respiratory quotient, body weight and energy usage. In addition, gene expression related to the Wnt signalling pathway was assessed [66]. ApcMin/+ mice were given Bifidobacterium bifidum (Bf) and Lactobacillus gasseri (Lg) probiotic strains, with a concentration of 10 [57] colony-forming units (CFU)/100 g of food or a combination of both probiotic strains together with microencapsulated quercetin at a dosage of 15 mg/100 g of food, for 73 days. Subsequently, energy metabolism, alterations in organ and body weight, histology of colon tissue, composition of intestinal microbiota and gene expression related to the Wnt signalling system were also assessed. The data showed that the microencapsulated supplement, consisting of probiotics and quercetin, effectively inhibited the advancement of colorectal cancer in ApcMin/+ mice [66].

Furthermore, an investigation conducted by Liu and Zhi in 2021 employed a rat model to create constipation, a recognized risk factor for colorectal issues, by providing loperamide. Subsequently, the impact of quercetin on constipation induced by loperamide was examined. The data indicated that administration of quercetin at doses of 25 and 50 mg/kg in rats resulted in an elevation in intestinal transit rate, as well as an increase in the concentration of short-chain fatty acids and levels of gastrin, motilin and substance P. Besides, quercetin enhanced the motility of the intestines and decreased the levels of somatostatin. The levels of aquaporin 3 (AQP3), transient receptor potential vanilloid 1 (TRPV1), glial cell line-derived neurotrophic factor (GDNF), enteric nerve-related factors, nitric oxide (NO) synthase, stem cell factor (SCF) and its receptor c-Kit were measured using Western blotting and reverse transcription-quantitative polymerase chain reaction (RT-qPCR). Quercetin administration was discovered to decrease loperamide-induced constipation by enhancing the levels of interstitial cells of Cajal indicators, such as stem cell factor, its receptor c-Kit and AQP3. Ultimately, their findings demonstrated that quercetin exhibited a safeguarding impact against loperamide-induced constipation, hence potentially mitigating the likelihood of developing colorectal cancer [67].

The antiangiogenic and anticancer effects of quercetin and luteolin on colon cancer cells (HT29) were assessed by Erdoğan et al. in comparison with the conventional chemotherapy drug 5-FU and a combination of 5-FU and luteolin or quercetin. An MTT (3-[4,5-dimethylthiazol-2-yl]-2,5-diphenyl tetrazolium bromide) assay, fluorescence microscopy, human VEGF ELISA (enzyme-linked immunosorbent assay), Western blotting and qRT-PCR were employed. Their method of evaluating the effects was Western blotting on the following genes: mTOR, Akt protein, PTEN, p53, Bax, Bcl-2 and P38. The MTT assay assessed cell viability, and ELISA was utilized to ascertain how the treatment affected angiogenesis. Fluorescence microscopy was employed to identify HT-29 cell apoptosis. Eight times as much apoptosis was induced in cells treated with quercetin and 5-FU as in cells treated with luteolin and 5-FU, and 10 times more in cells treated with the combination. The VEGF level exhibited a notable decrease in cells subjected to a combination treatment of quercetin or luteolin along with 5-FU. The researchers discovered that quercetin and luteolin can govern programmed cell death (apoptosis) in HT-29 cells. Additionally, it was observed that combination therapy decreased the expression of anti-apoptotic proteins, such as Bcl-2, mTOR and Akt genes, as compared to the control group. The groups treated with 5-FU and quercetin exhibited a more rapid rise in the expression of P53, P38, MPK and PTEN genes compared to those treated with 5-FU and luteolin. To summarize, the

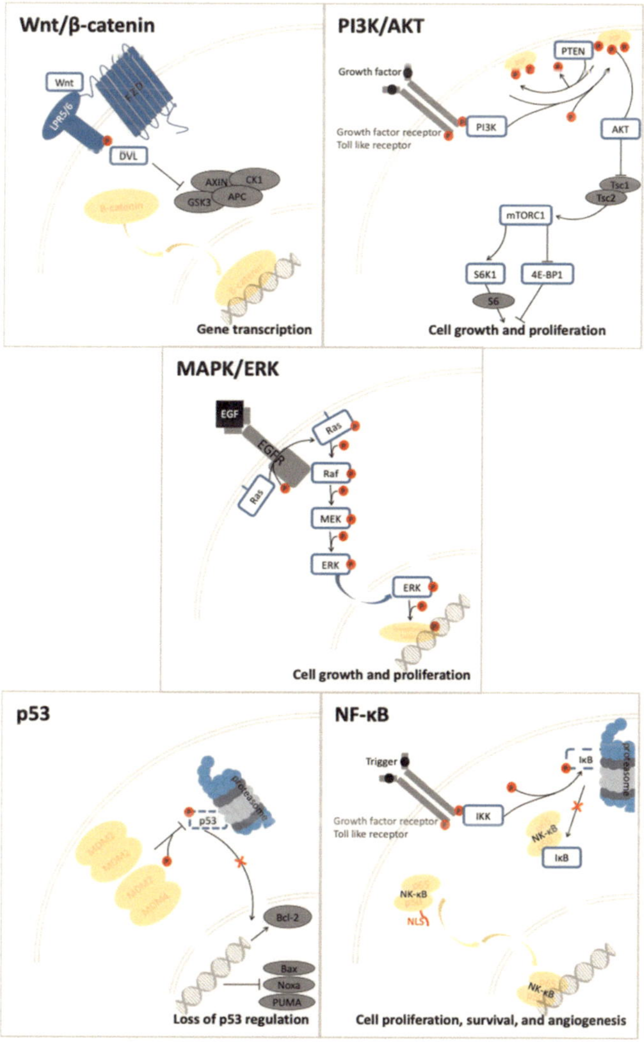

Figure 4.
Signalling transduction pathway in colorectal cancer that being modulated by therapeutic approaches of quercetin. Source: Adaptation by Ref. [69] under creative common license. https://creativecommons.org/licenses/by/4.0/

combined use of luteolin and quercetin can enhance the anticancer properties of 5-FU and mitigate its harmful side effects in colorectal cancer [68]. **Figure 4** shows signalling transduction pathways in colon commonly modulated by quercetin.

Gastrointestinal cancer	Preclinical model (*in vitro*/*in vivo*)	Anticancer mode	Molecular mechanism	Reference
Colorectal cancer	*In vivo* (rat)	Anti-apoptotic, anti-hyperproliferation	Quercetin reduced tumour occurrence and number by controlling β-catenin and APC expression. Histology confirmed quercetin's beneficial effect in reducing DMH-induced pathogenic changes.	[65]
	In vivo (ApcMin/+ mice)	—	Probiotics and quercetin-containing microencapsulated supplementation successfully prevented colorectal cancer in ApcMin/+ mice from progressing via alteration of Wnt signalling	[66]
	In vivo (rat)	—	Results showed that quercetin had a protective effect against constipation brought on by loperamide, which may lessen the risk of colorectal cancer by increasing intestinal transit rate and motility and reducing somatostatin levels	[67]
	In vitro (HT-29 cells)	Anti-angiogenic and anti-apoptotic	Treatment with 5-FU and quercetin increased expression levels of P53, P38, MPK and PTEN genes	[68]
	In vitro (HCT-116 Cells)	Anti-proliferation, inhibition on cell cycle and suppression of growth	Halted cell division at G1 phase	[69]
	In vivo	Apoptosis, anti-angiogenesis and anti-proliferation	Induced apoptosis, anti-angiogenesis and anti-proliferation via EGFR, Akt, Cdk1, cyclin B and VEGF alteration	[63]
	In vitro (Caco-2 cells)	Apoptosis and antioxidant	Apoptosis induction via TNF-α and TNF-R1 alteration	[71]
	In vitro (DLD-1 cells)	Apoptosis	Apoptosis induction via KRAS, JNK and caspase-3 alteration	[72]
	In vitro (Caco-2 cells)	Lipid peroxidation	Inhibition of lipid peroxidation via Nrf-2 and Prx-6 protein	[73]

Table 3
Summary on the latest studies on the therapeutic effect of quercetin in colorectal cancer.

Al-Ghamdi et al. examined the mechanism and impact of a 200 mg dosage of quercetin and a 150 mg dose of EGCG (epigallocatechin gallate) at different ratios on the stimulation of programmed cell death and suppression of growth in the human colorectal cell line HCT-116 [70]. Quercetin suppressed cell proliferation, halted the cell cycle, induced annexin V and decreased clonogenicity. The lowest dose of the investigated medicines suppressed colony development. Moreover, a substantial rise in annexin V was observed at 150 mg of quercetin and 100 mg of EGCG. The combination therapy resulted in the cessation of cell division at the G1 phase. Ultimately, the combination of EGCG and quercetin holds promise as a potent combination of chemotherapy in the future. However, further investigations are necessary to determine the appropriate dosage and potential adverse reactions [70]. **Table 3** lists some studies on the therapeutic effect of quercetin in colorectal cancer.

4.4 Pancreatic cancer

The plant-derived medication presents a compelling option for treating cancer. Quercetin has been demonstrated that quercetin modulates several dysregulated signalling pathways, including those linked to autophagy and apoptosis, in order to achieve its anticancer effects through a variety of methods. In particular, quercetin inhibits multiple signalling pathways, including NF-κB, P53, Wnt/β-catenin, MAPK, janus kinase/signal transducer and activator of transcription (JAK/STAT) and the Hedgehog pathway, to achieve its anticancer actions. Numerous intracellular signalling molecules, including tumour necrosis factor-α (TNF-α), Bax, Bcl-2, caspases and VEGF, are interfered by quercetin.

Numerous cancer types, including breast, prostate, ovarian, lung, colon, hepatocellular, lymphoma and pancreatic cancer, have been researched in relation to quercetin's anticancer properties. Nevertheless, most current research on quercetin's anticancer effects focus on cancer in laboratory animals.

According to a recent study on the impact of quercetin on pancreatic cancer, quercetin treatment altered the expression of 105 miRNAs, including the lethal-7 (let-7), miR-194, miR-103, miR-29, miR-125 and miR-106 families. These miRNAs are essential for inhibiting cell death and promoting invasion, proliferation and metastasis. Among these, let-7c is one of the most significant miRNAs. This miRNA controls Notch inhibitor Numbl, a downstream target of let-7c, following transcription. By intensifying its hostility towards Notch, the Numbl stops the spread of cancer [74].

Quercetin is one of the natural plant extracts that have various important biological antitumour properties. Besides, quercetin has been shown to exert its anticancer properties by targeting noncoding RNAs (ncRNAs) that include long non-coding RNA (lncRNA) and microRNA (miRNA), which have been linked to the development of cancer. It inhibits the proliferation of cancer cells, encourages cell death and increases susceptibility to chemotherapy drugs by controlling the production of ncRNAs, which in turn influences the expression of genes and proteins involved in the signalling cascade. These provide insight into the molecular relationship between quercetin and non-coding RNAs, which may be useful in using quercetin as a therapeutic adjuvant. When combined, quercetin is a naturally occurring substance that may have an anticancer impact when used as a cancer adjuvant [9].

Quercetin specifically causes death in pancreatic cancer cells to have an anticancer impact. Still, new research also suggests that quercetin modifies many signal transduction pathways to slow down the growth of cancer. Quercetin may prevent EMT, invasion and metastasis by suppressing the expression of the N-cadherin, MMP-9 and STAT-3 signalling pathways. Through its inhibitory influence on Receptor for Advanced Glycation Endproducts (RAGE) expression, quercetin increases the chemosensitivity of pancreatic cancer cells to gemcitabine (GEM). Meanwhile, it is a desirable agent for the treatment of cancer due to its broad accessibility, effectiveness and low toxicity when compared to other investigated substances. Recently, quercetin has been made available and used as a potentially effective medication to treat a variety of malignancies, either by itself or in conjunction with other chemotherapeutic drugs [75].

Guo et al. employed various methods such as a nude mice tumour formation test, migration, proliferation and invasion as well as real-time cell analysis, to assess the effects of quercetin on pancreatic ductal adenocarcinoma (PDA). The development and spread of tumours as well as the *in vivo* and *in vitro* flow cytometry investigation of Sonic hedgehog (SHH) signalling in pancreatic cancer cells and colony formation was examined [76]. Through the downregulation of c-Myc, quercetin demonstrated anticancer efficacy by inhibiting pancreatic growth. Quercetin inhibited migration and invasion by lowering transforming growth factor β1 (TGF-β1) levels and suppressing the epithelial-mesenchymal transition (EMT). Through the mitochondrial and death receptor pathways, quercetin caused apoptosis. In an investigation conducted on naked mice, quercetin therapy decreased the number of metastases. Quercetin's therapeutic actions on PDA are connected to Gli2 (GLI Family Zinc Finger 2) but not Gli1 (GLI Family Zinc Finger 1), which entails the modulation of SHH activity. Recombinant Shh protein was used to boost SHH activity, reversing quercetin's effects on pancreatic cancer cells' proliferation, migration and invasion. Furthermore, by upregulating Zeb2 and Snail1 expression, Shh may trigger TGF-β1/Smad2/3 signalling and induce EMT, partially reversing the inhibition of cancer cell migration and invasion that quercetin mediated. According to these results, quercetin may be utilized to treat PDA by preventing pancreatic cancer cells' invasion, migration and metastasis as well as by causing apoptosis by blocking the SHH and TGF-β1/Smad signalling pathways [77].

Hassan et al. assessed the effectiveness of quercetin in conjunction with well-known medications such as gemcitabine (GEM) and doxorubicin (DOX) against human pancreatic cancer cells and hepatic cancer cells, respectively. The investigation revealed that quercetin combined anticancer medications produced superior outcomes than single-drug therapy. Up to 60% of the cells in two-dimensional (2D) and three-dimensional (3D) cultures were dead, with single-drug treatment only exerting ≤20% apoptosis [78]. A more thorough analysis showed that quercetin downregulated hypoxia-inducible factor (HIF)-1α and raised p53 regulator levels, which led to p53-mediated apoptosis. The impact of quercetin on hypoxia and drug efflux pump function was also assessed by investigating the functions of HIF-1α and multidrug transporters (such as multidrug resistance (MDR)) activity [78]. The investigation revealed that quercetin could reduce multidrug resistance mutation 1 (MDR-1) efflux activity, exerting similar inhibitory effects as verapamil, the positive control used in the study. Hence, quercetin might be used with GEM or DOX against multidrug-resistant malignancies in the pancreas and liver, respectively [78].

In another investigation, Liu et al. assessed quercetin's effects and mode of action on pancreatic cancer cells resistant to GEM. Two cell lines from hepatocellular and pancreatic cancers were used in this investigation. Liver cancer cell lines used were HepG2 and Huh-7, whereas BxPC-3 and PANC-1 were the pancreatic cancer cell lines. Quercetin was reported to have a cytotoxic effect on HepG2 and PANC-1 (GEM-resistant) using a proliferation test and a pro-apoptotic effect on HepG2 and PANC-1, according to a flow cytometry analysis. Western blotting results showed that quercetin caused apoptosis by upregulating p53 protein and downregulating cyclin-D1. Additionally, it resulted in S phase cell cycle arrest. Ultimately, the findings demonstrated that quercetin can be used with established anticancer medications to treat GEM-resistant liver and pancreatic cancer [79].

Hoca et al. investigated the effects of quercetin and resveratrol on the epithelial-mesenchymal transition (EMT) in CD133-positive and CD133-negative pancreatic cancer cells [80]. After separating the CD133+ cancer stem cells from PANC-1 cells using the MiniMACS technique, quercetin and resveratrol were added in varying amounts to the three cell variants: CD133+, CD133− and PANC-1. The investigation employed MTT test and immunocytochemistry with antibodies against vimentin, TNF-α, actin alpha 2 (ACTA-2), N-cadherin and interleukin-1β (IL-1β). The intensity of N-cadherin, ACTA-2 and IL-1β staining was lower in the quercetin-treated cells compared to CD133 + cells treated with resveratrol. Hence, quercetin is effective in significantly reducing N-cadherin expression, preventing EMT which leads to metastasis [80].

In another investigation by Serri et al., gemcitabine and quercetin combination showed synergistic activity, particularly in inhibiting the migration of cancer cells. The study employs hyaluronic acid-coated nanoparticles (PPHA NPs) carrying GEM and quercetin [81]. The nanoparticle carriers were able to improve drug delivery, thereby improving cytotoxicity and cellular uptake; it was also mentioned that the PPHA NPs improved the anti-inflammatory properties of quercetin, observed by a

Gastrointestinal cancer	Preclinical model (*in vitro*/ *in vivo*)	Anticancer mode	Molecular mechanism	Reference
Pancreatic cancer	*In vivo* and *in vitro* (PANC-1 and Patu8988)	Inhibition of metastasis	EMT suppression by reducing TGF-*β*1 level, inhibition of growth, invasion and migration of cells, apoptosis of cancer cells by antagonizing TGF-*β*/Smad and SHH signalling pathways	[76]
	In vitro (Mia-PaCa-2 and PANC-1)	Synergism between gemcitabine and quercetin	Reduced IL-6 and IL-8 expressions and enhanced cytotoxicity against Mia-PaCa-2 and PANC-1 cell lines	[81]
	In vitro (PANC-1)	Inhibition of metastasis	Reduced immunoreactivities, such as ACTA-2, IL-1*β* and N-cadherin, increased TNF-*α* and vimentin, prevention of EMT	[82]
	In vivo and *in vitro* (PDAC)	Halting tumour progression	Improved effects of BET inhibitors at suppressing tumour development and reduced hnRNPA1 *in vivo*	[83]
	In vitro (MIA Paca-2, BxPC-3, AsPC-1, HPAC and PANC-1)	Apoptosis induction, autophagy, chemosensitizing effect	Quercetin showed a RAGE silencing like effect that attenuates RAGE expression to accelerate apoptosis, autophagy and chemosensitivity of MIA Paca-2^{GEMR} cells	[84]

Table 4
Anticancer effects of quercetin against pancreatic cancer.

reduction in interleukin (IL) levels in pancreatic ductal adenocarcinoma cell lines Mia-PaCa-2 and PANC-1 [81].

A few research on quercetin's potential therapeutic benefits for pancreatic cancer are included in **Table 4**.

4.5 Oesophageal cancer

Yue Liu et al. evaluated the effects of quercetin on the angiogenesis and migration of oesophageal cancer cells, in addition to the underlying mechanism [85]. In their study, human oesophageal cancer cells (Eca109) received 5 or 10 μg/ml of quercetin. A scratch wound healing assay evaluated cell migration and invasion was examined using a transwell assay, and a colony formation assay was conducted. Human umbilical vein vascular endothelial cells (CRL-1730) were inoculated in Eca109 conditioned medium, and the effects of quercetin were measured by tube formation and wound healing assays. Western blotting measured MMP-2, MMP-9 and VEGF-A protein expression levels. Quercetin (10 μg/ml) decreased colony formation in Eca109 cells; however, no difference was observed between the control group and the 5 μg/ml quercetin group. The group treated with 10 μg/ml quercetin showed reduced cell migration and invasion, while 5 μg/ml only suppressed invasion. Tube formation ability and endothelial cell migration were inhibited in cells incubated in Eca109 conditioned medium. The group treated with 10 μg/ml quercetin showed reduced MMP-2, MMP-9 and VEGF-A expression levels [85].

Interesting study by Li et al. involving Yishen Qutong Granules (YSQTG) in traditional Chinese medicine showed that quercetin (the main component of YSQTQ) could inhibit proliferation, invasion and clonal formation of oesophageal carcinoma cell lines [86]. Compared with the control group, quercetin at different concentrations of 50, 100, 150 and 200 μmol/L inhibited the proliferation of KYSE30 and KYSE150 cells at 24, 48 and 72 hours. Further study on the effects of quercetin on the formation of cell clonal ability showed that the number of clones formed in KYSE30 cells in the control and 50 and 100 μmol/L quercetin groups was 346 ± 7, 203 ± 10 and 132 ± 5, respectively, while the number of clones formed in KYSE150 cells in the control and 50 and 100 μmol/L quercetin groups was 470 ± 13, 364 ± 5 and 225 ± 15, respectively, the difference was statistically significant ($p < 0.05$). Molecular docking of quercetin and NF-κB was carried out and it was found that the binding energy of quercetin and NF-κB was −5.82 kcal/mol, indicating that small molecular compounds such as quercetin could directly bind to NF-κB stably. The specific binding of quercetin to NF-κB was confirmed, and the regulatory relationship of quercetin to NF-κB protein was also detected. The results showed that 50 and 100 μmol/L quercetin intervention groups could significantly reduce the level of NF-κB protein in KYSE150 cells ($p < 0.01$), suggesting that quercetin can inhibit the proliferation of oesophageal cancer cells in a concentration-dependent manner and directly target the key protein of NF-κB [86].

According to a study by Wang et al., quercetin treatment significantly slowed down the proliferation of oesophageal cancer cells by increasing the rate of apoptosis shown in the flow cytometry detection test ($p < 0.05$). Moreover, quercetin induced the expression of gene controlling Bax protein while suppressing the expression of Bcl-2 protein, further confirming its apoptosis-inducing effects on oesophageal cancer cells [87]. Combined with the results of MTT assays, it was shown that the treatment of quercetin substantially impaired the survival of oesophageal cancer cells ($p < 0.05$). The transwell assay showed that quercetin also inhibited the invasive capacity of oesophageal cancer cells, as evidenced by substantially fewer oesophageal cancer

Gastrointestinal cancer	Preclinical model (*in vitro*/*in vivo*)	Anticancer mode	Molecular mechanism	Reference
Oesophageal cancer	*In vitro* (Eca109 cells)	Inhibition of invasion and angiogenesis	Reduction levels of MMP-2, MMP-9 and VEGF-A	[85]
	In vitro (of KYSE30 and KYSE150 cells)	Inhibition of proliferation, invasion and clonal formation	affect the metabolic pathway, MAPK signalling pathway and nuclear factor-kappa B (NF-κB) signalling pathway	[86]
	In vitro (KYSE510 and TE-7 cells)	Induction of apoptosis	Inducing miR-1-3p level and suppressing TAGLN2 level	[87]

Table 5
Molecular mechanisms of quercetin on oesophageal cancer.

cells with quercetin treatments penetrating the membrane in the transwell chamber ($p < 0.05$) compared to normal oesophageal cancer cells. However, regarding the migration of oesophageal cancer cells detected by scratch assays, neither treatment of quercetin nor treatment of cisplatin showed significant impacts demonstrated by a similar closure rate, implying that the quercetin treatment majorly influences the invasion ability of cancer cells but had little effect on the migration. At the molecular level, the expression of miR-1-3p was induced ($p < 0.05$), while the expression of transgelin 2 (TAGLN2) was suppressed by quercetin ($p < 0.05$). Moreover, the anti-oesophageal cancer effects of quercetin were blocked by miR-1-3p inhibition ($p < 0.05$), which was represented by the restored growth and invasion of oesophageal cancer cells, demonstrating that quercetin exerted inhibitory effects on oesophageal cancer cells by inducing miR-1-3p [87]. **Table 5** summarizes the molecular mechanisms of quercetin in oesophageal cancer.

5. Conclusion and future perspectives

The use of quercetin in anticancer management has been proved to have a plethora of applications. Based on the above-mentioned evidences, quercetin makes a promising anticancer agent by targeting cancer cell proliferation, metastasis, angiogenesis and apoptosis directly on cancers of the GI tract. Furthermore, quercetin also prevents cancer progression by exerting chemopreventive effects via anti-oxidative properties. Various *in vitro* and *in vivo* studies have been cited to highlight the capabilities of quercetin in exerting anticancer effects, which include apoptosis induction via ROS activation, caspase-3 activation and downstream gene alteration. Quercetin also targets various molecular targets commonly involved in cancer, such as Akt, MAPK, epidermal growth factor receptor (EGFR), NF-κB and TNF-α, which abrogate oncogenic signalling. The safety profile of quercetin has also been described to be well tolerable in patients, with good bioavailability in glycoside form. Furthermore, quercetin has also shown great potential to be used in combination with other antineoplastic drugs, such as 5-FU, gemcitabine and doxorubicin, in combating cancer progression due to the synergism demonstrated. Hence, adjuvant therapy using quercetin can be a promising look into the near future, as it can improve drug efficiency and targeting as well as reduce adverse effects caused by high doses of anticancer drugs. Other fields of study that are worth looking into include interactions of quercetin with ncRNAs such as miRNA and lncRNA, as these fields are still left largely unexplored. These goals can be achieved with further mechanistic and safety studies, as well as taking the step further to include research on other cancers such as cancers of the central nervous system (CNS) and clinical studies in patients.

Acknowledgements

This work was supported by the Deanship of Scientific Research, Vice Presidency of Graduate Studies and Scientific Research, King Faisal University, Saudi Arabia (GrantA389) and Sunway University research grants (project no. GRTIN-IGS (02)-CVVR-13-2023).

Author details

Bernadette Xin Jie Tune[1], Yuan Seng Wu[2,3*], Rhanye Mac Guad[4], Aimi Syamima Abd Manap[5], Sheryar Afzal[5*], Kalaivani Batumalaie[6], Ker Woon Choy[7,8], Neeraj Kumar Fuloria[9], Shivkanya Fuloria[9], Vetriselvan Subramaniyan[10,11] and Mahendran Sekar[12]

1 Faculty of Pharmacy, Department of Pharmaceutical Life Sciences, University of Malaya, Kuala Lumpur, Malaysia

2 Department of Medical Education, School of Medical and Life Sciences, Sunway University, Selangor, Malaysia

3 Sunway Microbiome Centre, School of Medical and Life Sciences, Sunway University, Selangor, Malaysia

4 Faculty of Medicine and Health Science, Department of Biomedical Science and Therapeutics, Universiti Malaysia Sabah, Kota Kinabalu, Malaysia

5 Department of Biomedical Science, College of Veterinary Medicine, King Faisal University, Ahsaa, Saudi Arabia

6 Pre-University Programmes, Sunway College Johor Bahru, Johor Bahru, Johor, Malaysia

7 Institute of Pathology, Laboratory and Forensic Medicine (I-PPerForM), Universiti Teknologi MARA, Sungai Buloh, Malaysia

8 Department of Anatomy, Faculty of Medicine, Universiti Teknologi MARA (UiTM), Sungai Buloh, Malaysia

9 Faculty of Pharmacy, AIMST University, Bedong, Malaysia

10 Pharmacology Unit, Jeffrey Cheah School of Medicine and Health Sciences, Monash University, Subang Jaya, Malaysia

11 Center for Transdisciplinary Research, Department of Pharmacology, Saveetha Dental college, Saveetha Institute of Medical and Technical Sciences, Saveetha University, Chennai, Tamil Nadu, India

12 School of Pharmacy, Monash University Malaysia, Subang Jaya, Malaysia

*Address all correspondence to: sengwu_21@yahoo.com and safzal@kfu.edu.sa

References

[1] Ferlay J et al. Estimating the global cancer incidence and mortality in 2018: GLOBOCAN sources and methods. International Journal of Cancer. 2019; **144**(8):1941-1953. DOI: 10.1002/IJC.31937

[2] Bray F, Ferlay J, Soerjomataram I, Siegel RL, Torre LA, Jemal A. Global cancer statistics 2018: GLOBOCAN estimates of incidence and mortality worldwide for 36 cancers in 185 countries. CA: A Cancer Journal for Clinicians. 2018;**68**(6):394-424. DOI: 10.3322/CAAC.21492

[3] Sung H et al. Global cancer statistics 2020: GLOBOCAN estimates of incidence and mortality worldwide for 36 cancers in 185 countries. CA: A Cancer Journal for Clinicians. 2021; **71**(3):209-249. DOI: 10.3322/CAAC.21660

[4] Baskar R, Lee KA, Yeo R, Yeoh KW. Cancer and radiation therapy: Current advances and future directions. International Journal of Medical Sciences. 2012;**9**(3):193-199. DOI: 10.7150/IJMS.3635

[5] Hossain MS et al. Colorectal cancer: A review of carcinogenesis, global epidemiology, current challenges, risk factors, preventive and treatment strategies. Cancers (Basel). 2022;**14**(7): 1732. DOI: 10.3390/CANCERS14071732

[6] Ekor M. The growing use of herbal medicines: Issues relating to adverse reactions and challenges in monitoring safety. Frontiers in Pharmacology. 2014; **4**:177. DOI: 10.3389/FPHAR.2013.00177

[7] Anand David AV, Arulmoli R, Parasuraman S. Overviews of biological importance of quercetin: A bioactive flavonoid. Pharmacognosy Reviews. 2016;**10**(20):84-89. DOI: 10.4103/0973-7847.194044

[8] Salehi B et al. Therapeutic potential of quercetin: New insights and perspectives for human health. ACS Omega. 2020; **5**(20):11849-11872. DOI: 10.1021/ACSOMEGA.0C01818

[9] Asgharian P et al. Potential mechanisms of quercetin in cancer prevention: Focus on cellular and molecular targets. Cancer Cell International. 2022;**22**(1):257. DOI: 10.1186/S12935-022-02677-W

[10] Niedzwiecki A, Roomi MW, Kalinovsky T, Rath M. Anticancer efficacy of polyphenols and their combinations. Nutrients. 2016;**8**(9):552. DOI: 10.3390/NU8090552

[11] Boots AW, Haenen GRMM, Bast A. Health effects of quercetin: From antioxidant to nutraceutical. European Journal of Pharmacology. 2008;**585**(2–3): 325-337. DOI: 10.1016/J.EJPHAR.2008.03.008

[12] Sharma P, Hu-Lieskovan S, Wargo JA, Ribas A. Primary, adaptive, and acquired resistance to cancer immunotherapy. Cell. 2017;**168**(4): 707-723. DOI: 10.1016/J.CELL.2017.01.017

[13] Yang D, Wang T, Long M, Li P. Quercetin: Its main pharmacological activity and potential application in clinical medicine. Oxidative Medicine and Cellular Longevity. 2020;**2020**: 8825387. DOI: 10.1155/2020/8825387

[14] Wang W et al. The biological activities, chemical stability, metabolism and delivery systems of quercetin: A review. Trends in Food Science and

Technology. 2016;**56**:21-38. DOI: 10.1016/J.TIFS.2016.07.004

[15] Kapoor MP, Moriwaki M, Timm D, Satomoto K, Minegawa K. Genotoxicity and mutagenicity evaluation of isoquercitrin-γ-cyclodextrin molecular inclusion complex using Ames test and a combined micronucleus and comet assay in rats. The Journal of Toxicological Sciences. 2022;**47**(6):221-235. DOI: 10.2131/JTS.47.221

[16] Liang X, Xia Z, Yan J, Wang Y, Xue S, Zhang X. Quercetin inhibits human sperm functions by reducing sperm [Ca2+]i and tyrosine phosphorylation. Pakistan Journal of Pharmaceutical Sciences. 2016;**29** (6 Suppl):2391-2396

[17] Ożarowski M et al. Pharmacological effect of quercetin in hypertension and its potential application in pregnancy-induced hypertension: Review of in vitro, in vivo, and clinical studies. Evidence-Based Complementary and Alternative Medicine. 2018;**2018**: 7421489. DOI: 10.1155/2018/7421489

[18] Di Pierro F et al. Potential clinical benefits of quercetin in the early stage of COVID-19: Results of a second, pilot, randomized, controlled and open-label clinical trial. International Journal of General Medicine. 2021;**14**:2807-2816. DOI: 10.2147/IJGM.S318949

[19] Han MK, Barreto TA, Martinez FJ, Comstock AT, Sajjan US. Randomised clinical trial to determine the safety of quercetin supplementation in patients with chronic obstructive pulmonary disease. BMJ Open Respiratory Research. 2020;7(1):e000392. DOI: 10.1136/BMJRESP-2018-000392

[20] El-Saber Batiha G et al. The pharmacological activity, biochemical properties, and pharmacokinetics of the major natural polyphenolic flavonoid: Quercetin. Food. 2020;**9**(3):374. DOI: 10.3390/FOODS9030374

[21] Almeida AF et al. Bioavailability of quercetin in humans with a focus on interindividual variation. Comprehensive Reviews in Food Science and Food Safety. 2018;**17**(3):714-731. DOI: 10.1111/1541-4337.12342

[22] Septembre-Malaterre A et al. Focus on the high therapeutic potentials of quercetin and its derivatives. Phytomedicine Plus : International Journal of Phytotherapy and Phytopharmacology. 2022;**2**(1):100220. DOI: 10.1016/J.PHYPLU.2022.100220

[23] Dabeek WM, Marra MV. Dietary Quercetin and Kaempferol: Bioavailability and potential cardiovascular-related bioactivity in humans. Nutrients. 2019;**11**(10):2288. DOI: 10.3390/NU11102288

[24] Mansour FR, Abdallah IA, Bedair A, Hamed M. Analytical methods for the determination of quercetin and quercetin glycosides in pharmaceuticals and biological samples. Critical Reviews in Analytical Chemistry. 2023:1-26. DOI: 10.1080/10408347.2023.2269421

[25] Lesjak M et al. Antioxidant and anti-inflammatory activities of quercetin and its derivatives. Journal of Functional Foods. 2018;**40**:68-75. DOI: 10.1016/J.JFF.2017.10.047

[26] Ezzati M, Yousefi B, Velaei K, Safa A. A review on anti-cancer properties of Quercetin in breast cancer. Life Sciences. 2020;**248**:117463. DOI: 10.1016/J.LFS.2020.117463

[27] Arnold M et al. Global burden of 5 major types of gastrointestinal cancer. Gastroenterology. 2020;**159**(1):335-349.

e15. DOI: 10.1053/J.GASTRO.2020.02.068

[28] Plummer M, Franceschi S, Vignat J, Forman D, De Martel C. Global burden of gastric cancer attributable to *Helicobacter pylori*. International Journal of Cancer. 2015;**136**(2):487-490. DOI: 10.1002/IJC.28999

[29] Kamangar F et al. Opposing risks of gastric cardia and noncardia gastric adenocarcinomas associated with *Helicobacter pylori* seropositivity. Journal of the National Cancer Institute. 2006; **98**(20):1445-1452. DOI: 10.1093/JNCI/DJJ393

[30] Khan SA, Thomas HC, Davidson BR, Taylor-Robinson SD. Cholangiocarcinoma. Lancet. 2005; **366**(9493):1303-1314. DOI: 10.1016/S0140-6736(05)67530-7

[31] Siegel RL et al. Colorectal cancer statistics, 2020. CA: A Cancer Journal for Clinicians. 2020;**70**(3):145-164. DOI: 10.3322/CAAC.21601

[32] McCormack VA et al. Informing etiologic research priorities for squamous cell esophageal cancer in Africa: A review of setting-specific exposures to known and putative risk factors. International Journal of Cancer. 2017;**140**(2):259-271. DOI: 10.1002/IJC.30292

[33] Mali SB. Cancer treatment: Role of natural products. Time to have a serious rethink. Oral Oncology Reports. 2023;**6**: 100040. DOI: 10.1016/J.OOR.2023.100040

[34] Reyes-Farias M, Carrasco-Pozo C. The anti-cancer effect of quercetin: Molecular implications in cancer metabolism. International Journal of Molecular Sciences. 2019;**20**(13):3177. DOI: 10.3390/IJMS20133177

[35] Neuwirthová J, Gál B, Smilek P, Urbánková P. Potential of the flavonoid quercetin to prevent and treat cancer—Current status of research. Klinická Onkologie. 2018;**31**(3):184-190. DOI: 10.14735/AMKO2018184

[36] Zeng Y, Shen Z, Gu W, Wu M. Bioinformatics analysis to identify action targets in NCI-N87 gastric cancer cells exposed to quercetin. Pharmaceutical Biology. 2018;**56**(1):393-398. DOI: 10.1080/13880209.2018.1493610

[37] Lei CS, Hou YC, Pai MH, Lin MT, Yeh SL. Effects of quercetin combined with anticancer drugs on metastasis-associated factors of gastric cancer cells: In vitro and in vivo studies. The Journal of Nutritional Biochemistry. 2018;**51**: 105-113. DOI: 10.1016/J.JNUTBIO.2017.09.011

[38] Li H, Chen C. Quercetin has antimetastatic effects on gastric cancer cells via the interruption of uPA/uPAR function by modulating NF-κb, PKC-δ, ERK1/2, and AMPKα. Integrative Cancer Therapies. 2018;**17**(2):511-523. DOI: 10.1177/1534735417696702

[39] Shang HS et al. Quercetin induced cell apoptosis and altered gene expression in AGS human gastric cancer cells. Environmental Toxicology. 2018; **33**(11):1168-1181. DOI: 10.1002/TOX.22623

[40] Hemati M, Haghiralsadat F, Jafary F, Moosavizadeh S, Moradi A. Targeting cell cycle protein in gastric cancer with CDC20siRNA and anticancer drugs (doxorubicin and quercetin) co-loaded cationic PEGylated nanoniosomes. International Journal of Nanomedicine. 2019;**14**:6575-6585. DOI: 10.2147/IJN.S211844

[41] Balogh J et al. Hepatocellular carcinoma: A review. Journal of

Hepatocellular Carcinoma. 2016;**3**:41-53. DOI: 10.2147/JHC.S61146

[42] Méndez-Blanco C, Fondevila F, García-Palomo A, González-Gallego J, Mauriz JL. Sorafenib resistance in hepatocarcinoma: Role of hypoxia-inducible factors. Experimental & Molecular Medicine. 2018;**50**(10):1-9. DOI: 10.1038/S12276-018-0159-1

[43] Fondevila F, Méndez-Blanco C, Fernández-Palanca P, González-Gallego J, Mauriz JL. Anti-tumoral activity of single and combined regorafenib treatments in preclinical models of liver and gastrointestinal cancers. Experimental & Molecular Medicine. 2019;**51**(9):1-15. DOI: 10.1038/S12276-019-0308-1

[44] Fernández-Palanca P, Fondevila F, Méndez-Blanco C, Tuñón MJ, González-Gallego J, Mauriz JL. Antitumor effects of quercetin in hepatocarcinoma in vitro and in vivo models: A systematic review. Nutrients. 2019;**11**(12):2875. DOI: 10.3390/NU11122875

[45] Yamada N, Matsushima-Nishiwaki R, Kozawa O. Quercetin suppresses the migration of hepatocellular carcinoma cells stimulated by hepatocyte growth factor or transforming growth factor-α: Attenuation of AKT signaling pathway. Archives of Biochemistry and Biophysics. 2020;**682**:108296. DOI: 10.1016/J.ABB.2020.108296

[46] Ding Y, Chen X, Wang B, Yu B, Ge J, Shi X. Quercetin suppresses the chymotrypsin-like activity of proteasome via inhibition of MEK1/ERK1/2 signaling pathway in hepatocellular carcinoma HepG2 cells. Canadian Journal of Physiology and Pharmacology. 2018;**96**(5):521-526. DOI: 10.1139/CJPP-2017-0655

[47] Hisaka T et al. Quercetin suppresses proliferation of liver cancer cell lines in vitro. Anticancer Research. 2020; **40**(8):4695-4700. DOI: 10.21873/ANTICANRES.14469

[48] Abdu S, Juaid N, Amin A, Moulay M, Miled N. Effects of sorafenib and quercetin alone or in combination in treating hepatocellular carcinoma: In vitro and in vivo approaches. Molecules. 2022;**27**(22).DOI: 10.3390/MOLECULES27228082

[49] Zou H, Zheng YF, Ge W, Wang SB, Mou XZ. Synergistic anti-tumour effects of quercetin and oncolytic adenovirus expressing TRAIL in human hepatocellular carcinoma. Scientific Reports. 2018;**8**(1):2182. DOI: 10.1038/S41598-018-20213-7

[50] Guo X et al. The construction of preclinical evidence for the treatment of liver fibrosis with quercetin: A systematic review and meta-analysis. Phytotherapy Research. 2022;**36**(10): 3774-3791. DOI: 10.1002/PTR.7569

[51] Wu H et al. Quercetin inhibits the proliferation of glycolysis-addicted HCC cells by reducing hexokinase 2 and Akt-mTOR pathway. Molecules. 2019;**24**(10): 1993. DOI: 10.3390/MOLECULES24101993

[52] Ji Y et al. Quercetin inhibits growth of hepatocellular carcinoma by apoptosis induction in part via autophagy stimulation in mice. The Journal of Nutritional Biochemistry. 2019;**69**: 108-119. DOI: 10.1016/J.JNUTBIO.2019.03.018

[53] Chen Z et al. Reversal effect of quercetin on multidrug resistance via FZD7/β-catenin pathway in hepatocellular carcinoma cells. Phytomedicine. 2018;**43**:37-45. DOI: 10.1016/J.PHYMED.2018.03.040

[54] Abdelmoneem MA et al. Dual-targeted lactoferrin shell-oily core

nanocapsules for synergistic targeted/herbal therapy of hepatocellular carcinoma. ACS Applied Materials & Interfaces. 2019;**11**(30):26731-26744. DOI: 10.1021/ACSAMI.9B10164

[55] Ren KW et al. Quercetin nanoparticles display antitumor activity via proliferation inhibition and apoptosis induction in liver cancer cells. International Journal of Oncology. 2017; **50**(4):1299-1311. DOI: 10.3892/IJO.2017.3886

[56] Kovacovicova K et al. Senolytic Cocktail Dasatinib+Quercetin (D+Q) does not enhance the efficacy of senescence-inducing chemotherapy in liver cancer. Frontiers in Oncology. 2018;**8**:459. DOI: 10.3389/FONC.2018.00459

[57] Bhatiya M, Pathak S, Jothimani G, Duttaroy AK, Banerjee A. A comprehensive study on the anti-cancer effects of quercetin and its epigenetic modifications in arresting progression of colon cancer cell proliferation. Archivum Immunologiae et Therapiae Experimentalis (Warsz). 2023;**71**(1):6. DOI: 10.1007/S00005-023-00669-W

[58] Zhou W et al. Validation of quercetin in the treatment of colon cancer with diabetes via network pharmacology, molecular dynamics simulations, and in vitro experiments. Molecular Diversity. 2023. DOI: 10.1007/S11030-023-10725-4

[59] Inala MSR, Pamidimukkala K. Amalgamation of quercetin with anastrozole and capecitabine: A novel combination to treat breast and colon cancers—An in vitro study. Journal of Cancer Research and Therapeutics. 2023; **19**(Supplement):93-105. DOI: 10.4103/JCRT.JCRT_599_20

[60] Lawson MK. Improvement of therapeutic value of quercetin with chitosan nanoparticle delivery systems and potential applications. International Journal of Molecular Sciences. 2023;**24**(4):3293. DOI: 10.3390/IJMS24043293

[61] Tezerji S, abdolazimi H, Fallah A, Talaei B. The effect of resveratrol and quercetin intervention on azoxymethane-induced colon cancer in Rats model. Clinical Nutrition Open Science. 2022;**45**:91-102. DOI: 10.1016/J.NUTOS.2022.01.008

[62] Czapla J et al. The complex composition of trans-resveratrol, quercetin, vitamin E and selenium inhibits the growth of colorectal carcinoma. Anticancer Research. 2022; **42**(10):4763-4772. DOI: 10.21873/ANTICANRES.15981

[63] Rashedi J et al. Anti-tumor effect of quercetin loaded chitosan nanoparticles on induced colon cancer in Wistar rats. Advanced Pharmaceutical Bulletin. 2019;**9**(3):409-415. DOI: 10.15171/APB.2019.048

[64] Darband SG et al. Quercetin attenuated oxidative DNA damage through NRF2 signaling pathway in rats with DMH induced colon carcinogenesis. Life Sciences. 2020;**253**:117584. DOI: 10.1016/J.LFS.2020.117584

[65] Shree A, Islam J, Sultana S. Quercetin ameliorates reactive oxygen species generation, inflammation, mucus depletion, goblet disintegration, and tumor multiplicity in colon cancer: Probable role of adenomatous polyposis coli, β-catenin. Phytotherapy Research. 2021;**35**(4):2171-2184. DOI: 10.1002/PTR.6969

[66] Benito I et al. Microencapsulated *Bifidobacterium bifidum* and *Lactobacillus gasseri* in combination with quercetin inhibit colorectal cancer development in

ApcMin/+ mice. International Journal of Molecular Sciences. 2021;**22**(9):4906. DOI: 10.3390/IJMS22094906

[67] Liu W, Zhi A. The potential of Quercetin to protect against loperamide-induced constipation in rats. Food Science & Nutrition. 2021;**9**(6): 3297-3307. DOI: 10.1002/FSN3.2296

[68] Erdoğan MK, Ağca CA, Aşkın H. Quercetin and Luteolin improve the anticancer effects of 5-fluorouracil in human colorectal adenocarcinoma in vitro model: A mechanistic insight. Nutrition and Cancer. 2022;**74**(2): 660-676. DOI. 10.1080/ 01635581.2021.1900301

[69] Neamtu AA et al. A comprehensive view on the quercetin impact on colorectal cancer. Molecules. 2022;**27**(6): 1873. DOI: 10.3390/ MOLECULES27061873

[70] Al-Ghamdi MA, Al-Enazy A, Huwait EA, Albukhari A, Harakeh S, Moselhy SS. Enhancement of Annexin V in response to combination of epigallocatechin gallate and quercetin as a potent arrest the cell cycle of colorectal cancer. Brazilian Journal of Biology. 2021;**83**:e248746. DOI: 10.1590/ 1519-6984.248746

[71] Sezer ED, Oktay LM, Karadadaş E, Memmedov H, Selvi Gunel N, Sözmen E. Assessing anticancer potential of blueberry flavonoids, quercetin, kaempferol, and gentisic acid, through oxidative stress and apoptosis parameters on HCT-116 cells. Journal of Medicinal Food. 2019;**22**(11):1118-1126. DOI: 10.1089/JMF.2019.0098

[72] Yang Y et al. Quercetin preferentially induces apoptosis in KRAS-mutant colorectal cancer cells via JNK signaling pathways. Cell Biology International. 2019;**43**(2):117-124. DOI: 10.1002/ CBIN.11055

[73] Morales AM, Mukai R, Murota K, Terao J. Inhibitory effect of catecholic colonic metabolites of rutin on fatty acid hydroperoxide and hemoglobin dependent lipid peroxidation in Caco-2 cells. Journal of Clinical Biochemistry and Nutrition. 2018;**63**(3):175-180. DOI: 10.3164/JCBN.18-38

[74] Nwaeburu CC et al. Up-regulation of microRNA let-7c by quercetin inhibits pancreatic cancer progression by activation of Numbl. Oncotarget. 2016; 7(36):58367-58380. DOI: 10.18632/ ONCOTARGET.11122

[75] Asgharian P et al. Quercetin impact in pancreatic cancer: An overview on its therapeutic effects. Oxidative Medicine and Cellular Longevity. 2021;**2021**: 4393266. DOI: 10.1155/2021/4393266

[76] Guo Y et al. Quercetin suppresses pancreatic ductal adenocarcinoma progression via inhibition of SHH and TGF-β/Smad signaling pathways. Cell Biology and Toxicology. 2021;**37**(3): 479-496. DOI: 10.1007/S10565-020-09562-0

[77] Mirazimi SMA et al. Application of quercetin in the treatment of gastrointestinal cancers. Frontiers in Pharmacology. 2022;**13**:860209. DOI: 10.3389/FPHAR.2022.860209

[78] Hassan S et al. Quercetin potentializes the respective cytotoxic activity of gemcitabine or doxorubicin on 3D culture of AsPC-1 or HepG2 cells, through the inhibition of HIF-1α and MDR1. PLoS ONE. 2020;**15**(10): e0240676. DOI: 10.1371/JOURNAL. PONE.0240676

[79] Liu ZJ et al. Quercetin induces apoptosis and enhances gemcitabine

therapeutic efficacy against gemcitabine-resistant cancer cells. Anti-Cancer Drugs. 2020;**31**(7):684-692. DOI: 10.1097/CAD.0000000000000933

[80] Hoca M, Becer E, Kabadayı H, Yücecan S, Vatansever HS. The effect of resveratrol and quercetin on epithelial-mesenchymal transition in pancreatic cancer stem cell. Nutrition and Cancer. 2020;**72**(7):1231-1242. DOI: 10.1080/01635581.2019.1670853

[81] Serri C et al. Combination therapy for the treatment of pancreatic cancer through hyaluronic acid-decorated nanoparticles loaded with quercetin and gemcitabine: A preliminary in vitro study. Journal of Cellular Physiology. 2019;**234**(4):4959-4969. DOI: 10.1002/JCP.27297

[82] El-kott AF, Shati AA, Ali Al-kahtani M, Alharbi SA. The apoptotic effect of resveratrol in ovarian cancer cells is associated with downregulation of galectin-3 and stimulating miR-424-3p transcription. Journal of Food Biochemistry. 2019;**43**(12):e13072. DOI: 10.1111/JFBC.13072

[83] Pham TND et al. Quercetin enhances the anti-tumor effects of BET inhibitors by suppressing hnRNPA1. International Journal of Molecular Sciences. 2019;**20**(17):4293. DOI: 10.3390/IJMS20174293

[84] Lan CY, Chen SY, Kuo CW, Lu CC, Yen GC. Quercetin facilitates cell death and chemosensitivity through RAGE/PI3K/AKT/mTOR axis in human pancreatic cancer cells. Journal of Food and Drug Analysis. 2019;**27**(4):887-896. DOI: 10.1016/J.JFDA.2019.07.001

[85] Liu Y, Li CL, Xu QQ, Cheng D, Di Liu K, Sun ZQ. Quercetin inhibits invasion and angiogenesis of esophageal cancer cells. Pathology, Research and Practice. 2021;**222**:153455. DOI: 10.1016/J.PRP.2021.153455

[86] Li J et al. Network pharmacology and in vitro experimental verification on intervention of quercetin, present in Chinese medicine Yishen Qutong granules, on esophageal cancer. Chinese Journal of Integrative Medicine. 2023;**29**(3):233-243. DOI: 10.1007/S11655-022-3677-6

[87] Wang Y, Chen X, Li J, Xia C. Quercetin antagonizes esophagus cancer by modulating miR-1-3p/TAGLN2 pathway-dependent growth and metastasis. Nutrition and Cancer. 2022;**74**(5):1872-1881. DOI: 10.1080/01635581.2021.1972125

Chapter 9

Quercetin: A Flavonoid with Diverse Chemo Preventive Properties against Cancer

Mohammed I. Rushdi

Abstract

Quercetin, an exceptional and extraordinary flavonoid possessing bioactive properties, presents a plethora of benefits for the promotion of good health. The anti-tumor characteristics of quercetin have been well-documented in various in vitro and in vivo investigations, encompassing a wide range of cell lines and animal models. Quercetin, through the activation of caspase-3 and inhibition of the phosphorylation of Akt, mTOR, and ERK, as well as the reduction of β-catenin and stabilization of HIF-1α, augments apoptosis and autophagy in cancer. Additionally, quercetin curbs cancer cell metastasis by decreasing MMP and VEGF secretion. Significantly, the potent cytotoxicity of quercetin against cancer cells is accompanied by minimal or no adverse effects or harm to healthy cells.

Keywords: quercetin, flavonol, cytotoxicity, antiproliferative, anti-tumor, signaling pathway, non-coding RNAs

1. Introduction

As the second greatest cause of death worldwide, cancer is a complicated illness [1]. In actuality, the survival rate for cancer patients is still quite low despite advancements in diagnostic and treatment protocols, which include surgery, radiation, chemotherapy, and target and gene therapy [2]. The recurrence of cancer after months or years due to metastases—the spread of cancer cells from a primary lesion to distant sites—is most likely the true cause of the high cancer death rate [3]. The cancer and metastatic treatments that are now on the market usually cause substantial levels of toxicity and are linked to a variety of side effects that are frequently unanticipated and unexplained [4]. Therefore, it is highly recommended to develop new and more effective cancer treatment and metastasis prevention measures [5]. Phytochemicals are among the current class of therapeutic agents for cancer because of their low toxicity and capacity to target numerous cell signaling such as VEGF [6]. Quercetin, derived from quercetum (oak woodland) and called after Quercus, has been used since 1857. It is abundant in plants such as apples, berries, brassica vegetables, capers, grapes, onions, spring onions, tea, and tomatoes, as well as many seeds, nuts, flowers, bark, and leaves. However, quercetin is found in medicinal plants such as Ginkgo biloba, Hypericum perforatum, and elderberry and is mostly obtained from onions, apples, and tea. The molecule of quercetin has a ketocarbonyl group, and the oxygen atom on the first carbon is basic and can form salts with strong acids. It has four active groups in its molecular structure: a dihydroxy group between the A ring, o-dihydroxy group B, C ring C2, C3 double bond, and 4-carbonyl. The presence of a phenolic hydroxyl group and double bonds confers significant antioxidant action on quercetin. Its antioxidant and anti-inflammatory qualities have been linked to the prevention and treatment of cardiovascular disease and cancer. Furthermore, in vivo and in vitro investigations have revealed that quercetin has antibacterial activity and efficiently inhibits biofilm formation by blocking the expression of associated genes, antitumor activity, antiangiogenic activity, and so on. Furthermore, quercetin has a role [7]. Quercetin has numerous anti-inflammatory, antioxidant, and anticancer properties. Quercetin has been shown in both in vivo and in vitro studies to have anti-tumor actions by modifying cell cycle progression, reducing cell proliferation, inducing apoptosis, limiting angiogenesis and metastatic progression, and influencing autophagy. This study outlines the data regarding quercetin pharmacological potential and inhibition of malignancies, arguing that it should be investigated as a therapeutic agent against a variety of cancers [8]. The metabolite of quercetin, known as quercetin-3-glucuronide (Q3G), is one of the flavonoid classes present in fruits, vegetables, and medicinal plants. Due to its quick conjugation with glucuronic acid upon absorption from the small intestine, Q3G is also the main quercetin conjugate in human plasma, where quercetin is either missing or undetectable. Even though research has shown quercetin to have anti-viral, anti-microbial, anti-thrombotic, anti-carcinogenic, and neuroprotective properties, due to its short biological half-life, low water solubility, and chemical instability, it has poor bioavailability, which limits its usefulness in the food and pharmaceutical industries. Better than the previously listed, most quercetin in food and its metabolites are linked to a sugar molecule; this can boost bioavailability since sugar molecules are more soluble in water [9].

2. Therapeutic activity of quercetin in cancer

Numerous disorders have demonstrated the substantial positive effects of quercetin. More and more scientists are focusing on quercetin's potential as a tumor treatment because, at appropriate dosages, it does not appear to have any harmful side effects on normal cells. Quercetin significantly prevents the cell cycle, promotes apoptosis, and inhibits angiogenesis and metastasis in vitro, as demonstrated by some studies that have confirmed its ability to exert anti-tumor functions in a variety of mechanisms. These results have been confirmed in both in vitro and in vivo models of various tumors, with encouraging outcomes. According to the findings of in vivo research, quercetin at the chosen dosage effectively inhibits the growth of xenograft tumor models [8]. Moreover, quercetin has undergone clinical trials; these have revealed no toxicity or adverse effects on the general public [10]. Quercetin's capacity to reduce inflammation and lower the incidence of prostate cancer has been shown in clinical research [11, 12]. Phase I clinical trials demonstrated quercetin's anticancer efficacy and verified the drug's safety when administered intravenously. Quercetin-based clinical trials have demonstrated its ability to reduce blood pressure and improve anemia [13].

3. In vitro studies

3.1 Effect of quercetin on the cell cycle of cancer cells

The four main phases of the cell cycle are G1 (pre-DNA synthesis), S (DNA synthesis phase), G2 (late DNA synthesis phase), and M (mitotic phase). Cyclin, cyclin-dependent kinases (CDKs), and cyclin-dependent kinase inhibitors (CKI) are primarily responsible for controlling the cell cycle. The regulatory pathway of the pRb network primarily controls the cell cycle (**Figure 1**) [15].

Cyclin binds to its corresponding CDK to create a complex that phosphorylates Rb and releases cAb1 and E2F. E2F then reaches the nucleus to encourage the transition from the G1 phase to the S phase for cell-autonomous division. INK4 (which includes p15, p16, p18, and p19) is one type of CKI that competes with cyclin D1 for binding to CDK4/CDK6, which blocks the development of the cell cycle and phosphorylating Rb [16]. Consuming quercetin may be important for controlling the cell cycle, promoting apoptosis, and preventing metastasis. Moreover, quercetin inhibits the growth of colon cancer cells by modifying the expression of the anti-aging genes SIRT-6 and Klotho. Numerous molecular targets that may actively regulate cell proliferation, apoptosis, and cellular senescence were identified by the miRNA expression profile and functional enrichment analysis. Apart from the control of aging-associated miRNA and anti-aging protein activities like SIRT-6 and Klotho, telomerase activity suppression resulting in telomere length reduction established an intriguing path for the development of future colon cancer treatments. Therefore, a combination therapy combining quercetin and epigenetic modifiers in addition to currently available chemotherapeutic medications may assist in controlling the course of colon cancer and minimize its side effects [17]. To control cell cycle division, p53 primarily regulates another significant route that stimulates the expression of p21, GADD45, and Bax [18]. On the other hand, aberrant cyclin expression has been linked to aberrant cell cycle activity and unchecked tumor cell multiplication in research on tumor development. Quercetin has been demonstrated to induce cell cycle arrest at G2/M in human leukemia U937 cells. Cyclin D, E, and E2F levels decreased in tandem with the buildup of G2/M, while cyclin B levels significantly increased [19–22]. Comparable outcomes were also observed in ovarian cancer (SKOV3), where the cell was halted in the S and G2/M phases, and there was a decrease in cyclin D1 [23]. Furthermore, quercetin can alter the G0/G1 phase of HOS osteosarcoma cells and 232B4 chronic lymphocytic leukemia cells [24, 25]. Numerous research investigations have also examined the impact of quercetin on p53-related pathways and the advancement of malignant cell cycles. According to the study, quercetin can raise ER pressure, which in turn encourages p53's release. This inhibition of p53, CDK2, cyclin A, and cyclin B activity leads to the arrest of MCF-7 breast cancer cells at the S phase [26–29]. Quercetin caused MDA-MB-453 cells in a different study team's investigation of the breast cancer cell line to enter the G2/M phase arrest. Additionally, they discovered a noteworthy rise in p53 expression in their investigation [30, 31]. Quercetin can also impact the advancement of the cell cycle in other ways. A specific dose of quercetin-6-C-beta-d-glucopyranoside administration causes cell cycle arrest at the G0/G1 phase in the prostate cancer cell lines PC-3 and DU-145. Down-regulation of cyclin E and D, increased expression of p21 and p27, and protein expression of PNCA and CDK2 may be linked to this occurrence [32, 33]. Additionally, researchers discovered that 7-O-Geranylquercetin,

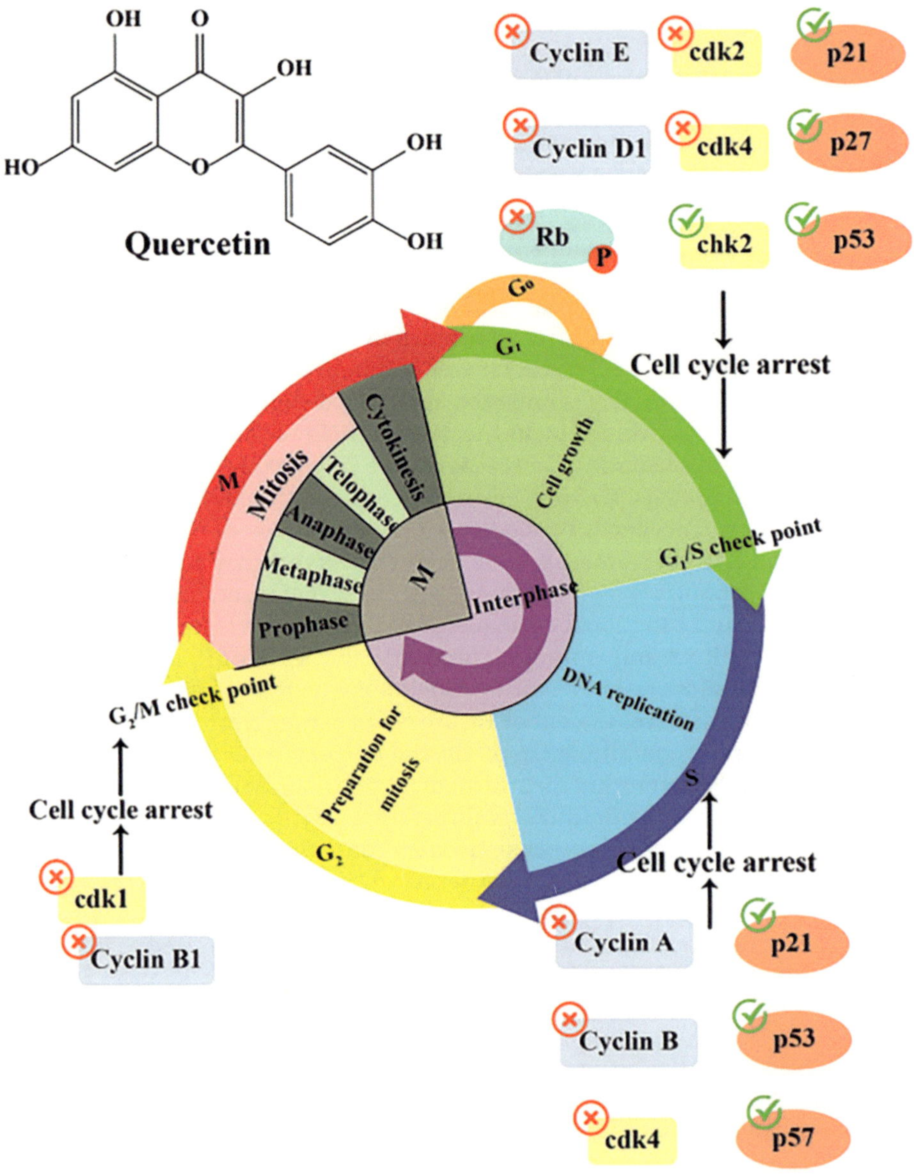

Figure 1.
How quercetin affects the cell cycle process by modulating several signaling molecules (√downregulation and ×upregulation by quercetin) [14]. Copyright Elsevier (2023).

a novel O-alkylated derivative of quercetin, could stop the G2/M cell cycle of MGC-803 and SGC-7901 gastric cancer cell lines [34, 35]. In YD10B and YD38 OSCC cells, quercetin reduced cell viability and caused G1 cell cycle arrest. Furthermore, quercetin significantly upregulated the expression of a CDK inhibitor while down-regulating the expression of proteins that promote cell cycle progression. In both kinds of OSCC cells, quercetin also markedly and dose-dependently boosted the number of annexin-V-positive cells. Because of quercetin apoptotic potential, PARP was cleaved, which activated the p38 MAPK-signaling pathway [21].

3.2 Effect of quercetin on apoptosis of cancer cells

The mechanism by which living organisms eliminate diseased or unnecessary cells from their bodies is widely recognized as apoptosis, or the process of cellular demise. It primarily encompasses two distinct apoptotic pathways, namely the extrinsic (also referred to as death receptor) and intrinsic (also known as mitochondrial) pathways (**Figure 2**) [36].

The extrinsic pathway is initiated by TRAILR and FAS death receptors, which collectively activate a cascade of proteases, including caspase-3, -6, -7, -8, -9, and -10, ultimately leading to cell death. On the other hand, the intrinsic pathway is triggered by the liberation of cytochrome c (cyt c) from the mitochondria. It is widely acknowledged that tumor cells possess the ability to evade death and proceed with tumor formation. Hence, the induction of apoptosis in tumor cells has become a significant breakthrough in the realm of anticancer therapy [37]. Quercetin has been shown to influence the apoptotic pathway and cause tumor cells to die. It has

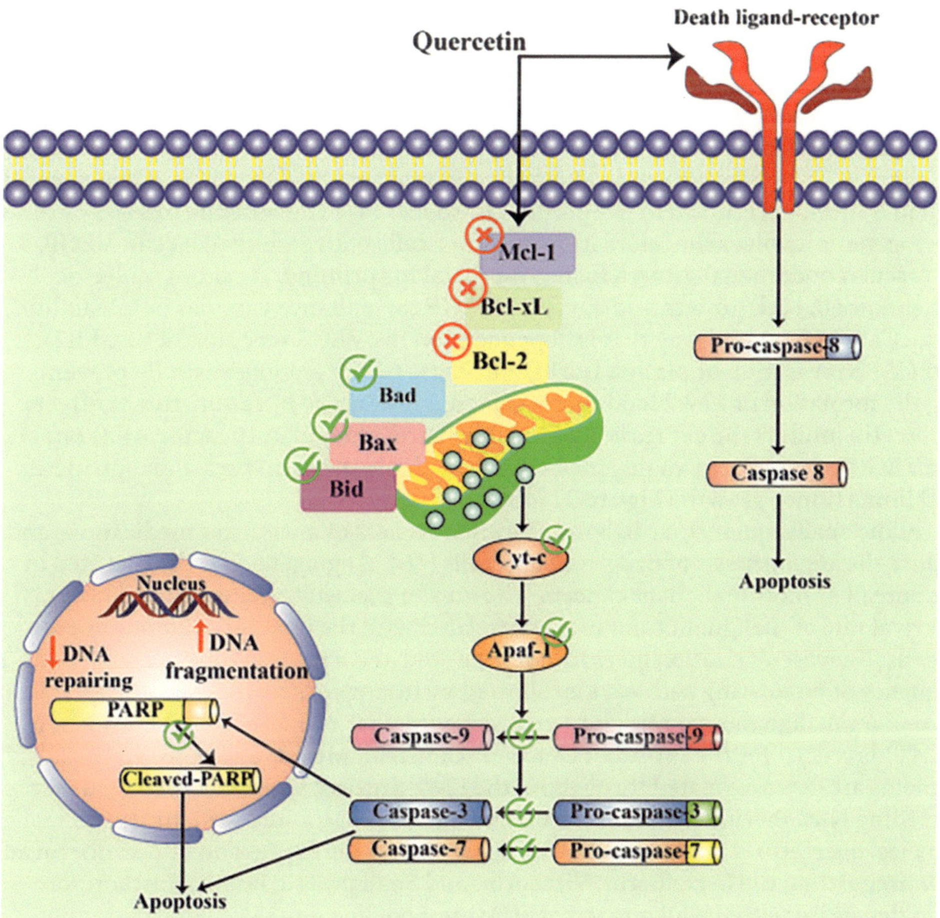

Figure 2.
Quercetin impact on the different apoptotic pathways (quercetin $^{\surd}$*downregulation and* ^{x}*overexpression, respectively) [14]. Copyright Elsevier (2023).*

been demonstrated that a suitable dosage of quercetin increases the expression of pro-apoptotic proteins and lowers the expression of anti-apoptotic proteins. A375SM melanoma cells [38], A2780S ovarian cancer cells, and HL-60 acute myeloid leukemia cells can all undergo apoptosis when treated with quercetin [39]. Cancer stem cells are essential to the disease's development. It has recently been established that quercetin can cause gastric cancer stem cells to undergo apoptosis [35, 40, 41]. Quercetin promotes the production of Bax and Bad during the intrinsic apoptotic pathway and downregulates the anti-apoptotic proteins Bcl-XL, Bcl-2, and Mcl-1. It also activates caspase-3, -8, and -9. Quercetin concurrently promoted cytochrome c release. Quercetin downregulated Akt, PLK-1, cyclin-B1, cyclin-A, CDC-2, CDK-2, and Bcl-2 while upregulating Bax, caspase-3, and p21. Furthermore, it has been observed to decrease STAT3 activation and promote STAT3 protein degradation in liver cancer cells [42]. Quercetin glycosides also have pro-apoptotic effects. Quercetin glycosides inhibited the expression of proteins linked to apoptosis and produced ROS and cyt c more quickly, all of which eventually caused cancer cells to undergo apoptosis [34]. In HepG2 cells, quercetin glycosides can also trigger caspase-3-induced apoptosis [43].

3.3 Effect of quercetin on angiogenesis and metastasis of cancer cells

The process of forming new capillaries is known as angiogenesis, and it is aided by endostatin, adhesion molecules, growth factors, and other substances. Physiological angiogenesis is linked to the development of the reproductive system and the healing of injuries. Tumor growth and metastasis are impacted by dysregulated angiogenesis, which is intimately linked to neoplastic disorders [44]. The intricate process of tumor angiogenesis involves the interaction of tumor cells with endothelial cells. VEGF, or vascular endothelial growth factor, is crucial for priming. A crucial molecule for endothelial cell growth and survival, VEGF can enhance vascular permeability, induce a signaling pathway that is dependent on the VEGF receptor 2 (VEGFR2), induce extravasation of plasma fibrin, and cause tumor angiogenesis. By preventing the formation of new blood vessels, quercetin can also have anti-tumor effects. Quercetin inhibits the expression of the downstream regulatory factor AKT, targets the VEGFR-2-mediated angiogenesis pathway in prostate and breast malignancies, and limits tumor growth (**Figure 3**) [45].

Additionally, quercetin can boost the effectiveness of anticancer medications and reduce the angiogenesis of drug-resistant cells [46]. Angiogenesis is the first step in the spread of most malignant cancers. The greater metastatic potential and lower survival rate of malignant tumors are correlated with the number of tumor microvessels. Neovascularization gives tumor cells food and a place to hide from the body. A sequence of circulating pathways is followed by tumor cells to infiltrate and establish secondary malignancies [47]. The process of epithelial-to-mesenchymal transition (EMT) is crucial for the spread of cancer. E-cadherin, MUC1, and other epithelial-type proteins are downregulated throughout the EMT process, while mesenchymal markers including N-cadherin, Vimentin, Snail, and others are acquired [48]. In several malignancies, quercetin can prevent EMT by upregulating the expression of E-cadherin and downregulating the N-cadherin, Vimentin, and Snail protein family. Furthermore, zinc-dependent extracellular matrix (ECM) remodeling de-endopeptidase family members known as matrix metalloproteinases (MMPs) are involved in various stages of cancer. Research has demonstrated that MMPs play a major role in the invasion and metastasis of cancer [49]. Quercetin can prevent tumor cell invasion and migration in both head and neck squamous cell carcinoma and colorectal cancer [50]. Quercetin

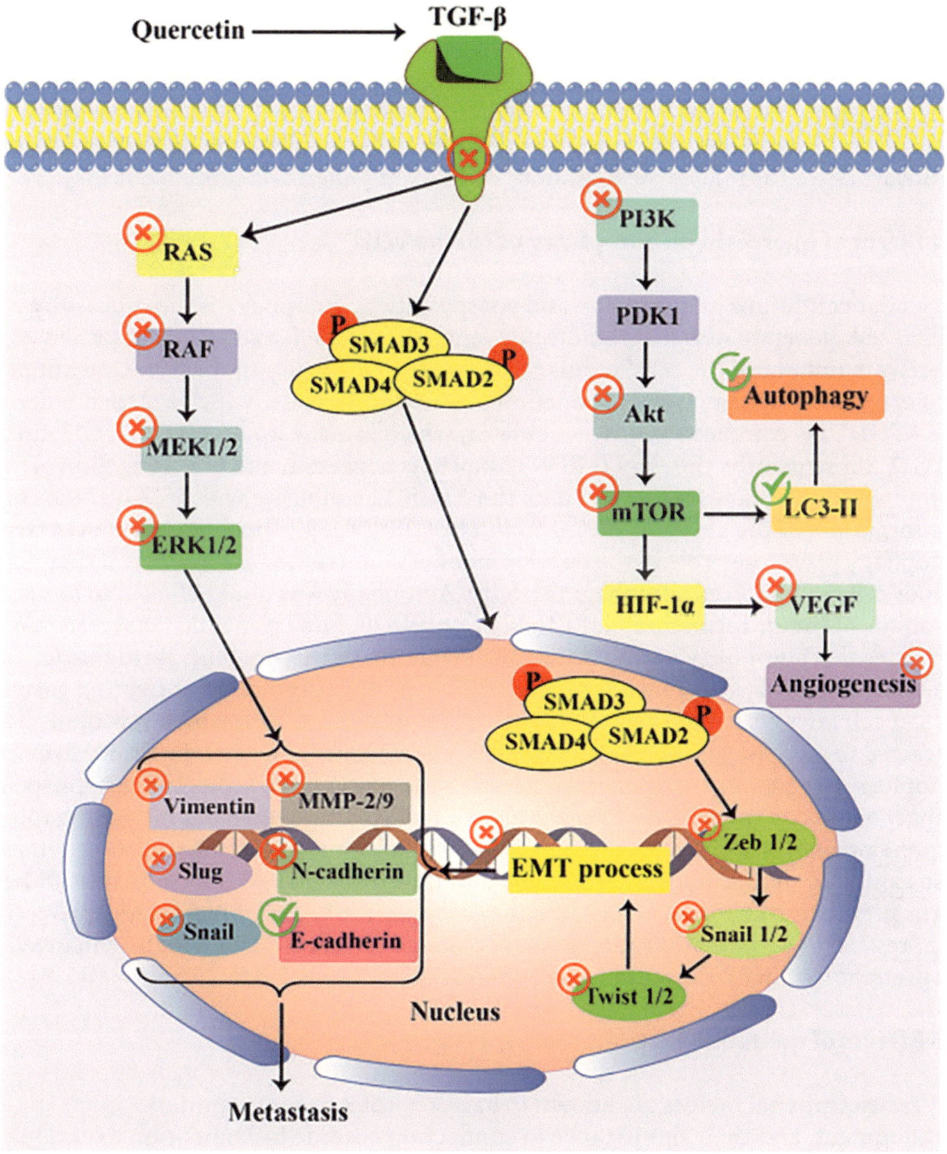

Figure 3.
Quercetin regulates the pathways involved in angiogenesis, autophagy, and metastasis (show quercetin respective effects on √downregulation and ˣ overexpression) [14]. Copyright Elsevier (2023).

blocks the Snail-dependent AKT activation pathway, which prevents lung cancer cells from invasively growing and metastasizing. N-cadherin, vimentin, ADAM9, and MMP-related protein expression were all markedly downregulated, whereas E-cadherin expression was markedly upregulated following quercetin administration [51–53]. Researchers employed TGF-β1 to generate EMT in cancer cells in colorectal cancer cell lines. To prevent EMT, quercetin administration can suppress Twist1 and control the expression of E-cadherin [54]. In PANC-1 and PATU-8988 pancreatic cancer cells, quercetin can reduce the expression of MMP2 and MMP7, which are intimately linked to the EMT process. It was also noted by researchers that quercetin can reverse the malignant progression brought on by IL-6 and prevent pancreatic cancer cells from invasively spreading by blocking the STAT3 signaling pathway [55]. By preventing c-Met's activation and the subsequent activation of Gabl, FAK, and PAK, quercetin can prevent HGF-induced melanoma cell migration [56]. Moreover, quercetin glycosides can prevent cancer from spreading. It can block the ERK1/2 and FAK pathways as well as reduce the migratory viability of pancreatic cancer cells [57].

3.4 Effect of quercetin on autophagy of cancer cells

Under conditions of starvation and energy deficit, autophagy is the process by which cells generate new macromolecules and ATP through a sequence of processes, thereby maintaining the cells' regular metabolism and ability to survive. One important stage of autophagy is the production of autophagosomes, which is dependent on the ATG1/ULK complex's positive regulator, which is made up of ATG1, ATG13, and ATG17. Subsequently, the class II PI3K complex is triggered, and the elongation of autophagic membranes is promoted by the ATG5-12 conjugate with 16. This leads to the formation of the autophagosome marker LC3II [58, 59]. Research has shown that autophagy plays a key role in the development of some tumor disorders and that many tumor cells have altered autophagic activity. Autophagy was once believed to be an inhibitor of tumor formation in the field of oncology; however, additional research indicates that autophagy may facilitate the evolution of tumors [60]. Autophagic vacuoles and acidic vesicular organelles (AVOs) were generated upon treating gastric cancer cell lines with quercetin. LC3I was transformed to LC3II, which was then attracted to autophagosomes to activate autophagy genes and start the protective autophagy progression in gastric cancer cells. Quercetin has a protective autophagous impact on gastric cancer cells by blocking AKT-mTOR signaling and increasing the expression of HIF-α [61]. Quercetin-treated glioblastoma and colon cancer cell lines also exhibited the accumulation of LC3II and the creation of AVOs, confirming quercetin stimulation of tumor cell protective autophagy [62, 63]. Nonetheless, quercetin suppression of proliferation and encouragement of apoptosis can both be enhanced by pretreatment with the autophagy inhibitor chloroquine [61].

3.5 Effects of quercetin on the transcription factors

Transcriptional factors are known to have key roles in tumor initiation and development, and their significance in connecting chronic inflammation to various cancers is well-established, even when it comes to distinct molecular targets [64]. Transcription is a highly regulated process in normal cells, and it is essential for normal cellular processes like differentiation and proliferation [65]. Conversely, signaling proteins that regulate distinct transcription factors are often dysfunctional in malignant cells and are thought to be the primary source of some oncogenic conversions in the function of these proteins, such as aberrant cell division and proliferation, anti-apoptosis, invasion, angiogenesis, metastasis, and resistance to chemotherapy and radiation [66]. Quercetin, like most phytochemicals, has been shown in numerous studies to target various cancer-associated molecules and pathways, including EGFR (epithelial growth factor receptor), ERK (extracellular signal-regulated kinase), MAPK (mitogen-activated protein kinase), NF-κB (nuclear factor kappa-light-chain-enhancer of activated B cells), STAT (signal transducer and activator of transcription), TGF-β (transforming growth factor-beta 1), TNF (tumor necrosis factor), Wnt/β-catenin (wingless-type MMTV integration site family), and so on.

Quercetin can be considered a promising anticancer agent [67]. Numerous studies have shown that dysregulation of various oncogenic transcription factors, including NF-κB, Wnt, Notch, MAPK, PI3K, and TGF, occurs in various cancer types, resulting in the induction, proliferation, and development of tumor cells. Consequently, these transcription molecules may be used to target malignancy.

3.5.1 Effects of quercetin on NF-κB

A transcription factor called NF-κB regulates the expression of several proteins involved in the control of important physiological processes, including angiogenesis (VEGF), apoptosis (Bcl-xL, cIAP1/2, FLICE, survivin, and XIAP), cell invasion (MMP-2, MMP-9, intercellular adhesion molecule-1 (ICAM-1), vascular cell adhesion protein-1 (ICAM-1), inflammation (interleukin-6 (IL-6), IL-1β, cyclooxygenase-2 (COX-2) and tumor necrosis factor (TNF)), metastasis (C-X-C chemokine receptor type-4 (CXCR)-4, TWIST), and proliferation (cyclin D1 and MYC) [68]. RelA (p65), RelB, c-Rel, NF-κB1/p105, and NF-κB2/p100 are the five primary transcription factors that make up NF-κB. These transcription factors are primarily inactive in the cytoplasm of most cells, but they can be triggered by a variety of substances, including growth factors, ROS, cytokines (IL-1β), and oncogenic stress [69, 70]. The stimulant parameters stated above activate the IKK complex's inhibitors, which are made up of three initial parts: IKK1/IKKα, IKK2/IKKβ, and NEMO (NF-κB essential modifier)/IKKγ. NF-κB activity is then regulated as a result [71]. When the IKK complex is activated, the IκB inhibitor proteins are phosphorylated, which leads to their destruction. Additionally, NF-κB signaling is activated to target gene expression [72]. The major signaling route that modulates NF-κB activation is PI3K/AKT. Quercetin (40 μM) treatment of PC-3 cell lines lacking p53 was shown to dramatically activate the PI3K/AKT pathway, leading to the phosphorylation of several molecules, including AKT (p-Thr308 and Ser473) and NF-κB-p105/p50 (p-Ser893) [22, 73–76].

3.5.2 Effects of quercetin on tyrosine kinase signaling

Two signaling pathways that contribute to the development of cancer include growth factor signaling and receptor tyrosine kinase (RTK) signaling. The human genome contains around 100 growth factors, of which approximately 90 genes encode proteins with tyrosine kinase activity. Furthermore, 58 of the 90 genes that make up the RTK are categorized into 20 distinct subfamilies. RTKs are cell surface receptors that control cell homeostasis; any malfunction can lead to a variety of clinical diseases. Five of the twenty groups listed— EGFR/ErbB, insulin receptors, PDGFR, FGFR, VEGFR, and CCK—are implicated in the initiation and spread of cancer [77]. RTKs are divided into three separate structural sections: the cytoplasmic, transmembrane, and extracellular regions. The extracellular portion of every cancer-related RTK is glycosylated, and the N-terminal region has many disulfide bonds. RTKs' extracellular portion possesses the ability to bind to ligands, perhaps causing TRK dimerization. RTKs are typically auto phosphorylated by trans-phosphorylating their kinase portion, which activates the cytoplasmic kinase portion of the protein and stabilizes by attaching to a dimeric ligand. The cytoplasmic portion consists of the juxtamembrane and C-terminal domains. The second one has tyrosine kinase, which catalyzes the phosphorylation of the receptor [78]. By reducing the expression of ErbB receptors, quercetin inhibited the growth and survival of prostate cancer cells that were androgen-dependent and -independent. In a 2005 study on colon cancer cell lines, it was shown that quercetin (0–100 μM) reduced the expression of ErbB2, ErbB3, and AKT signaling cascades in HT-29 cells, which in turn increased apoptosis [79, 80].

3.5.3 Effects of quercetin on Wnt signaling pathway

Nineteen genes in the human genome encode Wnt proteins, which control some human developmental events, such as cell division, death, migration, polarity, and destiny [14, 81]. The frizzled (FZD) receptors on Wnts—secreted glycoproteins rich in cysteines—identify them. LRP5 (LDL receptor-related proteins) or LRP6 are examples of phosphorylatable co-receptors that are necessary for the activation of FZDs, a class of seven transmembrane component surface cell receptors. These processes result in the complexation of FZDs with the cytoplasmic phosphor-protein disheveled (DVL). There are three known DVLs in humans: β-catenin, DVL1, and DVL2 [82]. A kind of basic Wnt, the Wnt/β-catenin pathway regulates several processes, including morphogenesis, embryogenesis, and the proliferation or differentiation of stem cells. Adenomatous polyposis coli (APC), a cytoplasmic multi-protein complex, destabilizes β-catenin to induce it to dissociate in the proteasomal processes in the absence of any stimuli. The Wnt pathway was inhibited by low intracellular concentrations of β-catenin through the transcription factor family TCF/LEF (T cell factor/lymphoid enhancer factor). β-catenin interaction with Wnt receptors is followed by suppressed subsequences of phosphorylation and degradation. Under these circumstances, the cytoplasmic presence of free β-catenin and its translocation into the nucleus cause the transcription of TCF/LEF-related genes, either stimulating or inhibiting apoptosis [83]. Treatment of prostate cancer (PC-3) cell lines with quercetin (0–60 μM) has been reported to upregulate E-cadherin and decrease TGF-β-triggered production of N-cadherin, Slug, Snail, Twist, and vimentin. Furthermore, quercetin decreased the development of prostate cancer by downregulating the two major Wnt signaling components, β-catenin and cyclin D1. This suggests that quercetin controlled the Wnt signaling axis and TGF-β-triggered EMT. Using A375 melanoma cells as a model, additionally, quercetin stimulated apoptosis in A375 cell lines by declining the regulation of Bcl-2 and activating caspase 3/7 via PARP cleavage [84].

3.5.4 Effects of quercetin on Notch signaling

The notch pathway, like the Wnt-β catenin cascade, controls multiple developmental pathways and defines the kind of stem cell differentiation. DLS (delta-serrate-Lag2) recognizes and activates Notch receptors as part of the notch pathway. After interacting with DLL (delta-like ligands) and JAG (jagged) DLS ligands, notch receptors become active [85]. Quercetin-ionizing radiation (IR) treatment of colon cancer stem cells (CSCs) had more anticancer effects than either IR or QU treatment alone. In xenograft mice-produced human colon cancer, quercetin-IR co-treatment (20 μM and 5 Gy, respectively) dramatically reduced CSC indicators and downregulated the expression of Notch-1 signaling molecules. Furthermore, in both DLD-1 and HT-29 cell lines, this co-treatment reduced the regulation of all five γ-secretase complex protein moieties (APH1, nicastrin, PEN2, presenilin-1, and presenilin-2). According to a different study, exposing breast cancer cell lines, such as MDA-MB-231, MCF-7, and T47D, to Quercetin-3-methyl ether (0–20 μM) decreased Notch-1 expression, which in turn prevented the EZH2 (enhancer of zeste homolog 2) pathway and PI3K and AKT phosphorylation, which in turn caused apoptosis [86].

3.5.5 Effects of quercetin on MAPK signaling

Differentiation, growth, migration, proliferation, apoptosis, and other cellular functions are all regulated by the mitogen-activated protein kinase (MAPK) pathway. Humans have been found to have six different types of MAPKs: ERK1/2, ERK3/4, ERK5, ERK7/8, JNK 1/2/3, and isoforms of p38 a/b/g (ERK6)/d [87]. MAPKs are thought to operate downstream of many membrane receptors, including the family of EGF receptors. Different MAPKs that primarily operate as transcription factors of genes are activated, and this regulates DNA replication, cell cycle progression, and division in addition to inflammatory and stress responses [88]. Quercetin (20 μM) treatment of MDA-MB-231 cells increased Foxo3a expression through pJNK/JNK upregulation, which in turn triggered cell cycle arrest and apoptosis [89].

3.5.6 Effects of quercetin on PI3K

The serine/threonine kinase group includes phosphatidylinositol 3-kinases (PI3Ks), which are triggered when growth factors interact with TRK. RTKs are phosphorylated at the tyrosine residue upon ligand binding. Through an adaptor protein, the phosphotyrosine can interact with PI3K directly or indirectly. Like MAPKs, PI3K activates a number of downstream kinases that significantly impact cell survival after activation [90]. Quercetin (0–200 μM) was found to promote apoptosis in leukemia cells by decreasing AKT phosphorylation and reducing the expression of PI3K and Bcl-2 proteins in an HL-60 cell research. Consequently, it reduced the viability and multiplication of leukemia cells. In an additional investigation, administering quercetin (0–200 μM) to DBTRG-05 and U-251 glioma cells resulted in a decrease in AKT protein levels. Additionally, quercetin Qu increased the effectiveness of temozolomide compared to conventional chemotherapy methods by blocking the PI3K/AKT signaling axis [91, 92].

3.5.7 Effects of quercetin on transforming growth factor-beta

The TGF-β family consists of around 40 members, which include Nodal, Activin, TGF-β, and bone morphogenetic proteins (BMGPs). TGF-β molecules are membrane proteins that carry out a variety of signaling functions from the cell membrane to the nucleus. This route encourages cell cycle arrest in the early stages of carcinogenesis, but as tumors grow, it also causes angiogenesis, invasion, and metastatic processes. Apart from SMAD-4, which plays a crucial role in modifying TGF-β's action, other signaling cascades that affect this one include MAPK, PI3K/AKT, and Wnt/β-catenin [93, 94]. The study conducted on colon cancer (SW-480 cells) showed that Qu (0–100 μM) inhibited TGF-β1-promoted EMT by upregulating E-cadherin through the downregulation of Twist1. Increased SHH expression in pancreatic cancer cells triggered the TGF-β1/Smad2/3 signaling pathway, which in turn promoted Snail and Zeb2 levels and EMT [94, 95].

3.6 Effects of quercetin on non-coding RNAs

3.6.1 Effects of quercetin on micro-RNAs (miRNAs)

Numerous studies have shown that one of the primary causes of breast cancer incidence is miRNA malfunction [96, 97]. Through the overexpression of p53, quercetin increased the expression of miR-34a, which in turn inhibited the function of SIRT1 (silent information regulator 1). As a result, increased p53 acetylation encouraged p53 stability and p53-associated apoptosis. Class III nuclear acetylase SIRT1 regulates multiple genomic processes through transcription factors and histone modification. It has been discovered that SIRT1 deacetylates a critical lysine residue, which dysregulates p53. Furthermore, miR-34a can inhibit SIRT1's mRNA by attaching to its 3′ UTR region [98, 99]. Quercetin may successfully stop the growth and invasion of lung cancer by targeting miRNAs. The expression of miRNAs in lung cancer cases was influenced by a quercetin-rich diet; the most notable miRNAs were oncogene miRNAs, such as the miR-17 and miR-146 families, and tumor suppressor miRNAs, such as the let-7 and miR-26 families. Integral membrane proteins of the claudin family regulate the functioning of tight junctions (TJs), and several cancer types have been linked to TJ malfunction [100]. A study conducted on lung adenocarcinoma A549 cell lines showed that quercetin (2.5–100 μM) promoted the expression of miR-16 in a time- and dose-dependent manner. This, in turn, reduced the stability of claudin-2 mRNA and slowed the growth of lung adenocarcinoma [101].

3.6.2 Effects of quercetin on long non-coding RNAs (lnc-RNAs) and circular RNAs (circ-RNAs)

It has been revealed that in NSCLC cells, miR-34a-5p downregulates lncRNA-SNHG7 (small nuclear RNA host gene 7), while lncRNA-SNHG7 regulation increased. Quercetin (0–100 μM) downregulated SNHG7, which led to the upregulation of miR-34a-5p. Therefore, quercetin inhibited the proliferation, viability, invasion, and migration of NSCLC cells by targeting quercetin/SNHG7/miR-34a-5p. One of the characteristics of prostate cancer is elevated expression of the lncRNA MALAT1. Research demonstrated that by modifying the regulation of EMT, PI3K/AKT, and apoptosis-associated molecules, quercetin (0–50 μM) reduced the expression of MALAT1, suppressing proliferation, growth, migration, and invasion in PC-3 cells [102]. Another family of single-strand, non-coding RNAs, known as circular RNAs (circRNAs), are produced from back-spliced exons of mRNAs and antisense RNAs. Because of their covalently closed structures, circRNAs are immune to RNAse R's breakdown. Applying Qu (0–200 μM) to HCT-116 colon cancer cells inhibited cell division by encouraging apoptosis. Thus, compared to untreated cells, quercetin-treated HCT-116 cells had differently regulated 240 lncRNA, 131 circRNA, 83 miRNA, and 1415 mRNA, according to the whole transcription sequencing technique. Furthermore, it was discovered that Quercetin inhibited leucine α-2-glycoprotein-1 (LRG1) regulation in HCT-116 cells to achieve its anticancer effects. Subsequent research revealed that Qu reduced the expression of LRG1 by downregulating the lncRNAs in a competitive manner. A different study using HeLa cancer cells showed that quercetin (0–50 μM) changed the regulation of approximately 10 miRNAs, 1 lncRNA (MALAT1), and 71 circRNAs [103].

3.7 In vivo studies

It has been demonstrated that quercetin inhibits the growth of several malignancies in different xenograft models. After receiving quercetin treatment, the tumor volume was drastically reduced, and the survival rate of animal models bearing tumors was greatly elevated. The encouragement of apoptosis and suppression of proliferation, angiogenesis, and metastasis are examples of how quercetin inhibits these processes in xenograft animal models [8]. In the models of breast cancer and leukemia cell xenograft, the inhibition of the AKT/mTOR pathway by varying doses of quercetin can induce apoptosis and impact the cell cycle [104]. Moreover, the effects of quercetin on tumor growth can be observed in vivo through the promotion of graft angiogenesis and metastasis. One of the earlier discovered anti-angiogenic factors, Thrombospondin-1 (TSP-1), has been found to inhibit tumor growth. A recent investigation has revealed that quercetin can upregulate the expression of TSP-1, leading to the inhibition of tumor growth in a prostate cancer model using mice [105]. Additionally, in a BALB/c mice model of breast cancer, Zhao et al. discovered that the administration of 34 mg/kg of quercetin can inhibit angiogenesis by targeting the calcineurin/NFAT pathway [106]. In vivo experiments have further substantiated the inhibitory impact of quercetin on the metastasis of tumors. Following the administration of 50 mg/kg quercetin, the incidence of colorectal lung metastasis was notably diminished [68]. The administration of quercetin has also been observed to deter epithelial-mesenchymal transition (EMT) by influencing the EGFR signaling pathway and curtailing the expression of VEGF [107]. Moreover, the suppressive effect of quercetin on the growth of allogeneic tumors has been identified in tumor cell models, including lung cancer and pancreatic cancer. Furthermore, quercetin exhibits the ability to synergize with other compounds in imparting anti-tumor activity [108]. For instance, the co-encapsulation of vincristine and quercetin within liposomal agents enhances the therapeutic efficacy of breast cancer treatment [109]. It was shown that the injection of MCF-7 cells into female BALB/c nude mice resulted in a decrease in the expression of VEGF, VEGFR, and NFATc3 in the tumor tissue under Qu (34 mg/kg/day) treatment. These findings revealed that quercetin targeted calcineurin to decrease angiogenesis in the MCF-7 cells of xenograft mice [106]. Additionally, it has been observed that quercetin (0–100 μM) decreased VEGFR expression in retinoblastoma cells (Y79) in a dose-dependent manner, indicating its anti-angiogenic effect [110].

4. Conclusions and perspectives

Quercetin, a kind of polyphenolic flavonoid, has anticancer properties. It has been demonstrated that quercetin exerts its anticancer effects by modulation of many dysregulated signaling pathways, including apoptosis and autophagy. Quercetin can stop the metastatic cascade at its start. It prevents extracellular matrix breakdown, tumor-associated angiogenesis, and Epithelial-mesenchymal transition. Except for a very small number of human clinical investigations that are not related to cancer, most quercetin-recognized properties have only been studied in vitro or in animal models. Furthermore, quercetin has not yet been the subject of any clinical trials as a cancer treatment. Quercetin displays anticancer actions via regulating multiple signaling pathways such as PI3K/AKT, NF-B, P53, Wnt/–catenin, MAPK, JAK/STAT, and Hedgehog. Quercetin inhibits many intracellular signaling molecules, including TNF-, Bax, Bcl-2, caspases, and VEGF. Quercetin has been researched for its anticancer properties in a variety of cancers, including breast cancer, prostate cancer, ovarian cancer, lung cancer, colon cancer, hepatocellular carcinoma, lymphoma, and pancreatic cancer. However, most new quercetin anticancer research is focused on cancer in humans. Quercetin has negligible adverse effects and negligible toxicity on normal cells while inducing death in malignant cells via both the intrinsic pathway and receptor-mediated extrinsic route by targeting various cellular signals. Furthermore, quercetin is an effective adjuvant to flavonoids, such as quercetin related-glycosylated derivatives, in addition to being able to overcome drug resistance.

Abbreviations

ABC	ATP-binding cassette
AIF	apoptosis-inducing factor-1
AMPK	adenosine monophosphate-activated protein kinase
BAK	BCL-2homologous antagonist killer
BAX	BCL-2 associated X protein
BCL-2	B-cell lymphoma-2
BIK	BCL-2 interacting killer
CCND	cyclin D
COX-2	cyclooxygenase-2
DNA	deoxyribonucleic acid
EGFR	epithelial growth factor receptor
EMT	endothelial to mesenchymal transition
ErbB	ErbB receptor tyrosine kinase 4
ERK	extracellular signal-regulated kinase
EZH2	enhancer of zeste homolog 2
H2AX	H2A histone family member X
HCC	hepatocellular carcinoma
HIF-1α	hypoxia-inducible factor-1α
hnRNP	heterogeneous nuclear ribonucleoprotein
IGF-R1	insulin-like growth factor type 1 receptor
IL	interleukin
JAK	Janus kinase
JNK	C-Jun N-terminal kinas
MALAT1	metastasis-associated lung adenocarcinoma transcript 1
MAPK	mitogen-activated protein kinase
MMP	matrix metalloproteinase
mTOR	mammalian target of rapamycin
ncRNA	non-coding RNA
NF-κB	nuclear factor kappa-light-chain-enhancer of activated B cells
PCNA	proliferating cell nuclear antigen
PI3K	phosphoinositide 3-kinases
Akt	protein kinase B
MUC4	Mucin 4
PKB	protein kinase B
PTEN	phosphatase and tensin homolog
PUMA	P53 upregulated modulator of apoptosis
ROS	reactive oxygen species
SIRT1	silent information regulator1
Sox2	sex-determining region Y-box2
c-Src	proto-oncogene tyrosine-protein kinase Src
STAT	signal transducer and activator of transcription
TGF-β	transforming growth factor beta 1
TNF	tumor necrosis factor
TWIST1	twist family BHLH transcription factor 1
VEGF	vascular endothelial growth factor
Wnt/β-catenin	wingless-type MMTV integration site family
XIAP	X-linked inhibitor of apoptosis

DOI: http://dx.doi.org/10.5772/intechopen.1004133

Author details

Mohammed I. Rushdi
Faculty of Pharmacy, Department of Pharmacognosy, South Valley University, Qena, Egypt

*Address all correspondence to: mrushdy258@svu.edu.eg

References

[1] Siegel RL et al. Cancer statistics, 2022. CA: A Cancer Journal for Clinicians. 2022;**72**(1):7-33

[2] Singh V, Khan N, Jayandharan GR. Vector engineering, strategies and targets in cancer gene therapy. Cancer Gene Therapy. 2022;**29**(5):402-417

[3] Kirkland JL. Tumor dormancy and disease recurrence. Cancer and Metastasis Reviews. 2023;**42**(1):9-12

[4] Pagani M et al. Hypersensitivity reactions to chemotherapy: An EAACI position paper. Allergy. 2022;**77**(2):388-403

[5] Gonzalez-Valdivieso J et al. Advanced nanomedicine and cancer: Challenges and opportunities in clinical translation. International Journal of Pharmaceutics. 2021;**599**:120438

[6] Attia EZ et al. In vitro cytotoxic potential of ethanol extract of Dictyopteris acrostichoides against human cancer cells. Brazilian Journal of Pharmacognosy. 2024;**34**:212-216

[7] Yang D et al. Quercetin: Its main pharmacological activity and potential application in clinical medicine. Oxidative Medicine and Cellular Longevity. 2020;**2020**:8825387

[8] Tang S-M et al. Pharmacological basis and new insights of quercetin action in respect to its anti-cancer effects. Biomedicine & Pharmacotherapy. 2020;**121**:109604

[9] Yu PR et al. The inhibitory effect of quercetin-3-glucuronide on pulmonary injury in vitro and in vivo. Journal of Food and Drug Analysis. 2023;**31**(2):254-277

[10] Al-Serwi RH et al. Targeted drug administration onto cancer cells using hyaluronic acid–quercetin-conjugated silver nanoparticles. Molecules. 2023;**28**(10):4146

[11] McCann SE et al. Intakes of selected nutrients, foods, and phytochemicals and prostate cancer risk in Western New York. Nutrition and Cancer. 2005;**53**(1):33-41

[12] O'Callaghan N, Douglas P, Keaver L. A qualitative study into cancer survivors' relationship with nutrition post-cancer treatment. Journal of Human Nutrition and Dietetics. 2023;**36**(2):406-414

[13] Kooshyar MM et al. A randomized placebo-controlled double blind clinical trial of quercetin in the prevention and treatment of chemotherapy-induced oral mucositis. Journal of Clinical and Diagnostic Research. 2017;**11**(3):Zc46-zc50

[14] Homayoonfal M et al. Quercetin modulates signal transductions and targets non-coding RNAs against cancer development. Cellular Signalling. 2023;**107**:110667

[15] Roelofs PA et al. Aberrant APOBEC3B expression in breast cancer is linked to proliferation and cell cycle phase. Cell. 2023;**12**(8):1185

[16] Ingham M, Schwartz GK. Cell-cycle therapeutics come of age. Journal of Clinical Oncology. 2017;**35**(25):2949-2959

[17] Bhatiya M et al. A comprehensive study on the anti-cancer effects of quercetin and its epigenetic modifications in arresting progression of colon cancer cell proliferation.

Archivum Immunologiae et Therapiae Experimentalis. 2023;**71**(1):6

[18] Otto T, Sicinski P. Cell cycle proteins as promising targets in cancer therapy. Nature Reviews Cancer. 2017;**17**(2):93-115

[19] Lee T-J et al. Quercetin arrests G2/M phase and induces caspase-dependent cell death in U937 cells. Cancer Letters. 2006;**240**(2):234-242

[20] Lomphithak T et al. Natural flavonoids quercetin and kaempferol targeting G2/M cell cycle-related genes and synergize with Smac mimetic LCL-161 to induce necroptosis in cholangiocarcinoma cells. Nutrients. 2023;**15**(14):3090

[21] Son HK, Kim D. Quercetin induces cell cycle arrest and apoptosis in YD10B and YD38 Oral squamous cell carcinoma cells. Asian Pacific Journal of Cancer Prevention. 2023;**24**(1):283-289

[22] Wang W et al. Quercetin induces MGMT+ glioblastoma cells apoptosis via dual inhibition of Wnt3a/β-catenin and Akt/NF-κB signaling pathways. Phytomedicine. 2023;**118**:154933

[23] Catanzaro D et al. Effect of quercetin on cell cycle and cyclin expression in ovarian carcinoma and osteosarcoma cell lines. Natural Product Communications. 2015;**10**(8):1934578X1501000813

[24] Suh DK et al. Induction of G1/S phase arrest and apoptosis by quercetin in human osteosarcoma cells. Archives of Pharmacal Research. 2010;**33**(5):781-785

[25] Hasan AA et al. Potentiation of cisplatin cytotoxicity in resistant ovarian cancer SKOV3/cisplatin cells by quercetin pre-treatment. International Journal of Molecular Sciences. 2023;**24**(13):10960

[26] Chou C-C et al. Quercetin-mediated cell cycle arrest and apoptosis involving activation of a caspase cascade through the mitochondrial pathway in human breast cancer MCF-7 cells. Archives of Pharmacal Research. 2010;**33**(8):1181-1191

[27] Niazvand F et al. Effects of quercetin-loaded nanoparticles on MCF-7 human breast cancer cells. Medicina. 2019;**55**(4):114

[28] Sabzini M et al. Development of chitosan/halloysite/graphitic-carbon nitride nanovehicle for targeted delivery of quercetin to enhance its limitation in cancer therapy: An in vitro cytotoxicity against MCF-7 cells. International Journal of Biological Macromolecules. 2023;**226**:159-171

[29] Dilek B, Meltem Ö. Quercetin suppresses cell proliferation using the apoptosis pathways in MCF-7 and MDA-MB-231 human breast carcinoma cells in monolayer and spheroid model cultures. South African Journal of Botany. 2023;**162**:259-270

[30] Choi EJ, Bae SM, Ahn WS. Antiproliferative effects of quercetin through cell cycle arrest and apoptosis in human breast cancer MDA-MB-453 cells. Archives of Pharmacal Research. 2008;**31**(10):1281-1285

[31] Tang H et al. Quercetin inhibits the metabolism of arachidonic acid by inhibiting the activity of CYP3A4, thereby inhibiting the progression of breast cancer. Molecular Medicine. 2023;**29**(1):127

[32] Hamidullah et al. Quercetin-6-C-β-d-glucopyranoside, natural analog of quercetin exhibits anti-prostate cancer activity by inhibiting Akt-mTOR pathway via aryl hydrocarbon receptor. Biochimie. 2015;**119**:68-79

[33] Lotfi N et al. The potential anti-cancer effects of quercetin on blood, prostate and lung cancers: An update. 2023;**14**:1-24

[34] Zhu Y et al. 7-O-Geranylquercetin induces apoptosis in gastric cancer cells via ROS-MAPK mediated mitochondrial signaling pathway activation. Biomedicine & Pharmacotherapy. 2017;**87**:527-538

[35] Rong Y et al. Quercetin inhibits the proliferative effect of gastric cancer cells by activating the pyroptosis pathway. Asian Journal of Surgery. 2023;**46**(11):5286-5288

[36] Banerjee N et al. Sequence driven interaction of amino acids in de-novo designed peptides determines c-Myc G-quadruplex unfolding inducing apoptosis in cancer cells. Biochimica et Biophysica Acta (BBA) - General Subjects. 2023;**1867**(2):130267

[37] Gourisankar S et al. Rewiring cancer drivers to activate apoptosis. Nature. 2023;**620**(7973):417-425

[38] Kim S-H et al. Antitumor and apoptotic effects of quercetin on human melanoma cells involving JNK/P38 MAPK signaling activation. European Journal of Pharmacology. 2019;**860**:172568

[39] Lee W-J et al. Quercetin induces mitochondrial-derived apoptosis via reactive oxygen species-mediated ERK activation in HL-60 leukemia cells and xenograft. Archives of Toxicology. 2015;**89**(7):1103-1117

[40] Shen X et al. Quercetin inhibits the growth of human gastric cancer stem cells by inducing mitochondrial-dependent apoptosis through the inhibition of PI3K/Akt signaling. International Journal of Molecular Medicine. 2016;**38**(2):619-626

[41] Hu M, Song H-Y, Chen L. Quercetin acts via the G3BP1/YWHAZ axis to inhibit glycolysis and proliferation in oral squamous cell carcinoma. Toxicology Mechanisms and Methods. 2023;**33**(2):141-150

[42] Sethi G et al. Apoptotic mechanisms of quercetin in liver cancer: Recent trends and advancements. Pharmaceutics. 2023;**15**(2):712

[43] Sudhanshu S, Vasantha RHP. Quercetin-3-O-glucoside induces human DNA topoisomerase II inhibition, cell cycle arrest and apoptosis in hepatocellular carcinoma cells. Anticancer Research. 2014;**34**(4):1691

[44] Al Amili M et al. Unforeseen consequences of synergizing PEGylated rapamycin and quercetin in angiogenesis and metastasis inhibition: Turning failure into benefit. Chemical Engineering Journal. 2023;**470**:144155

[45] Shahbaz M et al. Anticancer and apoptosis inducing potential of quercetin against a wide range of human malignancies. International Journal of Food Properties. 2023;**26**(1):2590-2626

[46] Lei C-S et al. Effects of quercetin combined with anticancer drugs on metastasis-associated factors of gastric cancer cells: In vitro and in vivo studies. Journal of Nutritional Biochemistry. 2018;**51**:105-113

[47] Rocchetti MT et al. Multi-faceted role of luteolin in cancer metastasis: EMT, angiogenesis, ECM degradation and apoptosis. International Journal of Molecular Sciences. 2023;**24**(10):8824

[48] Dai J et al. Chemoproteomics reveals baicalin activates hepatic CPT1 to ameliorate diet-induced obesity and hepatic steatosis. Proceedings of the National Academy of Sciences. 2018;**115**(26):E5896-E5905

[49] Yadav L et al. Matrix metalloproteinases and cancer-roles in threat and therapy. Asian Pacific Journal of Cancer Prevention. 2014;**15**(3):1085-1091

[50] Chan C-Y et al. Quercetin suppresses cellular migration and invasion in human head and neck squamous cell carcinoma (HNSCC). Biomedicine. 2016;**6**(3):15

[51] Chang J-H et al. Quercetin suppresses the metastatic ability of lung cancer through inhibiting snail-dependent Akt activation and snail-independent ADAM9 expression pathways. Biochimica et Biophysica Acta (BBA) - Molecular Cell Research. 2017;**1864**(10):1746-1758

[52] Zhou B et al. Quercetin inhibits DNA damage responses to induce apoptosis via SIRT5/PI3K/AKT pathway in non-small cell lung cancer. Biomedicine & Pharmacotherapy. 2023;**165**:115071

[53] Qin J, Fan H, Shi J. Quercetin induced biosynthesis of reduced graphene oxide and its application for treatment of lung cancer in nursing care. Inorganic Chemistry Communications. 2023;**155**:111036

[54] Feng J et al. Quercetin restrains TGF-β1-induced epithelial–mesenchymal transition by inhibiting Twist1 and regulating E-cadherin expression. Biochemical and Biophysical Research Communications. 2018;**498**(1):132-138

[55] Yu D et al. Quercetin inhibits epithelial–mesenchymal transition, decreases invasiveness and metastasis, and reverses IL-6 induced epithelial–mesenchymal transition, expression of MMP by inhibiting STAT3 signaling in pancreatic cancer cells. Oncotargets and Therapy. 2017;**10**(null):4719-4729

[56] Cao H-H et al. Quercetin inhibits HGF/c-Met signaling and HGF-stimulated melanoma cell migration and invasion. Molecular Cancer. 2015;**14**(1):103

[57] Lee J et al. Quercetin-3-O-glucoside suppresses pancreatic cancer cell migration induced by tumor-deteriorated growth factors in vitro. Oncology Reports. 2016;**35**(4):2473-2479

[58] Hurley JH, Young LN. Mechanisms of autophagy initiation. Annual Review of Biochemistry. 2017;**86**:225-244

[59] Vargas JNS et al. The mechanisms and roles of selective autophagy in mammals. Nature Reviews Molecular Cell Biology. 2023;**24**(3):167-185

[60] Levy JMM, Towers CG, Thorburn A. Targeting autophagy in cancer. Nature Reviews Cancer. 2017;**17**(9):528-542

[61] Wang K et al. Quercetin induces protective autophagy in gastric cancer cells: Involvement of Akt-mTOR-and hypoxia-induced factor 1α-mediated signaling. Autophagy. 2011;7(9):966-978

[62] Bi Y et al. Inhibition of autophagy induced by quercetin at a late stage enhances cytotoxic effects on glioma cells. Tumor Biology. 2016;**37**(3):3549-3560

[63] Kim H et al. Quercetin induces mitochondrial mediated apoptosis and protective autophagy in human glioblastoma U373MG cells. Oxidative Medicine and Cellular Longevity. 2013;**2013**:596496

[64] Aggarwal BB, Vijayalekshmi R, Sung B. Targeting inflammatory pathways for prevention and therapy of cancer: Short-term friend, long-term foe. Clinical Cancer Research. 2009;**15**(2):425-430

[65] Shanmugam MK, Kannaiyan R, Sethi G. Targeting cell signaling and

apoptotic pathways by dietary agents: Role in the prevention and treatment of cancer. Nutrition and Cancer. 2011;**63**(2):161-173

[66] Aggarwal BB et al. Targeting cell signaling pathways for drug discovery: An old lock needs a new key. Journal of Cellular Biochemistry. 2007;**102**(3):580-592

[67] Reyes-Farias M, Carrasco-Pozo C. The anti-cancer effect of quercetin: Molecular implications in cancer metabolism. International Journal of Molecular Sciences. 2019;**20**(13):3177

[68] Kee J-Y et al. Inhibitory effect of quercetin on colorectal lung metastasis through inducing apoptosis, and suppression of metastatic ability. Phytomedicine. 2016;**23**(13):1680-1690

[69] Sen R, Baltimore D. Multiple nuclear factors interact with the immunoglobulin enhancer sequences. Cell. 1986;**46**(5):705-716

[70] Matsumori A. Nuclear factor-κB is a prime candidate for the diagnosis and control of inflammatory cardiovascular disease. European Journal of Cardiology. 2023;**18**:e40

[71] Ahmad S et al. Long non-coding RNAs regulated NF-κB signaling in cancer metastasis: Micromanaging by not so small non-coding RNAs. Seminars in Cancer Biology. 2022;**85**:155-163

[72] Chauhan A et al. Phytochemicals targeting NF-κB signaling: Potential anti-cancer interventions. Journal of Pharmaceutical Analysis. 2022;**12**(3):394-405

[73] Ward AB et al. Quercetin inhibits prostate cancer by attenuating cell survival and inhibiting anti-apoptotic pathways. World Journal of Surgical Oncology. 2018;**16**(1):1-12

[74] Li L et al. Quercetin improves cerebral ischemia/reperfusion injury by promoting microglia/macrophages M2 polarization via regulating PI3K/Akt/NF-κB signaling pathway. Biomedicine & Pharmacotherapy. 2023;**168**:115653

[75] Fawzy MA et al. Quercetin potentiates the hepatoprotective effect of sildenafil and/or pentoxifylline against intrahepatic cholestasis: Role of Nrf2/ARE, TLR4/NF-κB, and NLRP3/IL-1β signaling pathways. Life Sciences. 2023;**314**:121343

[76] Lokman MS et al. Protective effect of quercetin against 5-fluorouracil-induced cardiac impairments through activating Nrf2 and inhibiting NF-κB and caspase-3 activities. Environmental Science and Pollution Research. 2023;**30**(7):17657-17669

[77] Du Z, Lovly CM. Mechanisms of receptor tyrosine kinase activation in cancer. Molecular Cancer. 2018;**17**(1):58

[78] Trenker R, Jura N. Receptor tyrosine kinase activation: From the ligand perspective. Current Opinion in Cell Biology. 2020;**63**:174-185

[79] Kim WK et al. Quercetin decreases the expression of ErbB2 and ErbB3 proteins in HT-29 human colon cancer cells. Journal of Nutritional Biochemistry. 2005;**16**(3):155-162

[80] García-Gutiérrez N et al. Quercetin and its fermented extract as a potential inhibitor of bisphenol A-exposed HT-29 colon cancer cells' viability. International Journal of Molecular Sciences. 2023;**24**(6):5604

[81] Buechling T, Boutros M. Chapter two - Wnt signaling: Signaling at and above the receptor level. In: Birchmeier C, editor. Current Topics in Developmental Biology. Elsevier Academic Press; 2011. pp. 21-53

[82] MacDonald BT, Tamai K, He X. Wnt/β-catenin signaling: Components, mechanisms, and diseases. Developmental Cell. 2009;**17**(1):9-26

[83] Ma Q et al. Wnt/β-catenin signaling pathway-a versatile player in apoptosis and autophagy. Biochimie. 2023;**211**:57-67

[84] Srivastava NS, Srivastava RAK. Curcumin and quercetin synergistically inhibit cancer cell proliferation in multiple cancer cells and modulate Wnt/β-catenin signaling and apoptotic pathways in A375 cells. Phytomedicine. 2019;**52**:117-128

[85] Meurette O, Mehlen P. Notch signaling in the tumor microenvironment. Cancer Cell. 2018;**34**(4):536-548

[86] Cao L et al. Quercetin-3-methyl ether suppresses human breast cancer stem cell formation by inhibiting the Notch1 and PI3K/Akt signaling pathways. International Journal of Molecular Medicine. 2018;**42**(3):1625-1636

[87] Rovida E, Stecca B. Mitogen-activated protein kinases and hedgehog-GLI signaling in cancer: A crosstalk providing therapeutic opportunities? Seminars in Cancer Biology. 2015;**35**:154-167

[88] Adewumi I, López C, Davie JR. Mitogen and stress-activated protein kinase regulated gene expression in cancer cells. Advances in Biological Regulation. 2019;**71**:147-155

[89] Nguyen LT et al. Quercetin induces apoptosis and cell cycle arrest in triple-negative breast cancer cells through modulation of Foxo3a activity. The Korean journal of physiology pharmacology. 2017;**21**(2):205

[90] Jiang N et al. Role of PI3K/AKT pathway in cancer: The framework of malignant behavior. Molecular Biology Reports. 2020;**47**(6):4587-4629

[91] Pozsgai E et al. Quercetin increases the efficacy of glioblastoma treatment compared to standard chemoradiotherapy by the suppression of PI-3-kinase-Akt pathway. Nutrition and Cancer. 2013;**65**(7):1059-1066

[92] Ahmadi M et al. Investigating the role of quercetin in increasing the rate of cisplatin-induced apoptosis via the NF-κB pathway in MG-63 cancer cells. Drug Research (Stuttgart). 2022;**72**(07):385-389

[93] Farooqi AA, Naureen H, Attar R. Regulation of cell signaling pathways by circular RNAs and microRNAs in different cancers: Spotlight on Wnt/β-catenin, JAK/STAT, TGF/SMAD, SHH/GLI, NOTCH and hippo pathways. Seminars in Cell & Developmental Biology. 2022;**124**:72-81

[94] Geng F et al. Quercetin alleviates pulmonary fibrosis in Silicotic mice by inhibiting macrophage transition and TGF-β-Smad2/3 pathway. Current Issues in Molecular Biology. 2023;**45**(4):3087-3101

[95] Guo Y et al. Quercetin suppresses pancreatic ductal adenocarcinoma progression via inhibition of SHH and TGF-β/Smad signaling pathways. Cell Biology and Toxicology. 2021;**37**(3):479-496

[96] Banerjee N et al. Cytotoxicity of pomegranate polyphenolics in breast cancer cells in vitro and vivo: Potential role of miRNA-27a and miRNA-155 in cell survival and inflammation. Breast Cancer Research and Treatment. 2012;**136**(1):21-34

[97] Zargar S et al. Evaluation of the possible pathways involved in

the protective effects of quercetin, naringenin, and Rutin at the gene, protein and miRNA levels using In-silico multidimensional data analysis. Molecules. 2023;**28**(13):4904

[98] Audrito V et al. Nicotinamide blocks proliferation and induces apoptosis of chronic lymphocytic leukemia cells through activation of the p53/miR-34a/SIRT1 tumor suppressor network. Cancer Research. 2011;**71**(13):4473-4483

[99] Asgharian P et al. Potential mechanisms of quercetin in cancer prevention: Focus on cellular and molecular targets. Cancer Cell International. 2022;**22**(1):257

[100] Ding L et al. The claudin family of proteins in human malignancy: A clinical perspective. Cancer Management Research. 2013;**5**:367-375

[101] Sonoki H et al. Quercetin decreases claudin-2 expression mediated by up-regulation of microRNA miR-16 in lung adenocarcinoma A549 cells. Nutrients. 2015;**7**(6):4578-4592

[102] Lu X et al. Quercetin inhibits epithelial-to-mesenchymal transition (EMT) process and promotes apoptosis in prostate cancer via downregulating lncRNA MALAT1. Cancer Management Research. 2020;**12**:1741-1750

[103] Li Y et al. Screening of therapeutic candidate genes of quercetin for cervical cancer and analysis of their regulatory network. OncoTargets Therapy. 2021;**14**:857-866

[104] Rivera Rivera A et al. Anti-breast cancer potential of quercetin via the Akt/AMPK/mammalian target of rapamycin (mTOR) signaling cascade. PLoS One. 2016;**11**(6):e0157251

[105] Yang F et al. Quercetin inhibits angiogenesis through thrombospondin-1 upregulation to antagonize human prostate cancer PC-3 cell growth in vitro and in vivo. Oncology Reports. 2016;**35**(3):1602-1610

[106] Zhao X et al. Quercetin inhibits angiogenesis by targeting calcineurin in the xenograft model of human breast cancer. European Journal of Pharmacology. 2016;**781**:60-68

[107] Jia L et al. Quercetin suppresses the mobility of breast cancer by suppressing glycolysis through Akt-mTOR pathway mediated autophagy induction. Life Sciences. 2018;**208**:123-130

[108] Wang P et al. Enhanced inhibition of prostate cancer xenograft tumor growth by combining quercetin and green tea. Journal of Nutritional Biochemistry. 2014;**25**(1):73-80

[109] Wong M-Y, Chiu GNC. Liposome formulation of co-encapsulated vincristine and quercetin enhanced antitumor activity in a trastuzumab-insensitive breast tumor xenograft model. Nanomedicine: Nanotechnology, Biology and Medicine. 2011;**7**(6):834-840

[110] Song W et al. Quercetin inhibits angiogenesis-mediated human retinoblastoma growth by targeting vascular endothelial growth factor receptor. Oncology Letters. 2017;**14**(3):3343-3348

Section 6

Cardiovascular Effects

Chapter 10

Quercetin as a Possible Cardiovascular Agent

Marek Pytliak and Viliam Vaník

Abstract

Diseases of the cardiovascular system are among the most common causes of morbidity and mortality in the adult population in developed countries. In addition to the possibilities of pharmacological treatment, the positive (and negative) influence of diet and its components is well documented in many cardiovascular diseases. Atherosclerosis is one of the main causes of chronic cardiovascular diseases. It is a chronic inflammatory disease of the vascular wall associated with disorders of lipid metabolism, endothelial dysfunction, migration, and proliferation of smooth muscle cells of the vascular media, oxidative stress, and many other mechanisms. Reactive oxygen species (ROS) contribute to the pathogenesis of many cardiovascular diseases. An imbalance between the formation of ROS and the ability of antioxidant systems to eliminate them leads to oxidative stress. Inhibition of ROS generation and function is thought to be a potential therapy to attenuate the extent of various cardiovascular diseases. The results of several studies indicate that the cardioprotective effect of quercetin could be related to its antioxidant properties. In the presented chapter, we will discuss the possible effects of quercetin on the prevention and treatment of various mechanisms supporting atherogenesis and thus the development of cardiovascular diseases.

Keywords: quercetin, prevention, antiischemic activity, cardiovascular disease, oxidative stress, free radicals

1. Introduction

The most common cause of death in the world is cardiovascular diseases (CVDs). It is estimated that about 17.9 million people worldwide die of cardiovascular diseases every year. In recent decades, CVDs have been the primary cause of death in developed countries, but currently, developing countries are rapidly approaching them [1]. The basic pathology of atherosclerotic blood vessel involvement results in the development of coronary artery diseases, peripheral vascular diseases, cerebrovascular diseases, and subsequent development of complications of these diseases. The main risk factors for the development and progression of atherosclerosis are relatively well known. For didactic reasons, atherosclerosis risk factors can be divided into influenceable and uninfluenceable. Uninfluenceable factors include, for example, male gender, age, and positive family history. Influenceable factors include arterial hypertension, lipid metabolism disorders, smoking, diabetes, abdominal obesity, lack of physical activity, and many other factors that can be influenced by non-pharmacological procedures or pharmacologically [2]. The preventive measures within the framework of cardiovascular diseases focus precisely on influencing those risk factors, either by non-pharmacological or pharmacological approaches. Flavonoids have recently been the subject of considerable interest by many experts due to their wide range of possible benefits. Due to their wide distribution in nature and their general availability, they have become the subject of many studies aimed at preventing and treating cardiovascular diseases [3]. In 1998, the first randomized clinical trial that investigated the effect of quercetin on cardiovascular health in healthy subjects was conducted [4]. One of the problems and a possible explanation for the inconsistent results of clinical trials with quercetin is its low bioavailability. This is mainly influenced by its low solubility in the gastrointestinal tract and rapid biodegradation into inactive metabolites. Moreover, there is probably a high inter-individual variability in the absorption of quercetin, glucosidase activity as well as variations in the performance of metabolic enzymes, which leads to different results even when using the same doses and forms of quercetin [5]. The increasing amount of evidence in favor of positive cardiovascular effects has supported the clinical relevance of modified dosage forms of quercetin using different carriers, such as nanoparticles and microemulsions, in the following years.

2. Pathogenesis of atherosclerosis—influence of oxidative stress

The development of individual stages of atherosclerosis is caused by many molecular and cellular events at each level—from an early lesion of fatty streaks to a highly dangerous plaque prone to rupture. As mentioned above, the production of ROS such as superoxide anion, hydrogen peroxide, lipid peroxides, and peroxynitrite and their dominance over antioxidant systems leads to oxidative stress, which, together with endothelial dysfunction, is one of the important processes in the initiation and progression of atherosclerosis. Dietary antioxidants (natural free radical scavengers) and enzymes, e.g., glutathione peroxidase, can inactivate various species of ROS. Therefore, the low concentrations of these substances act as proatherosclerotic factors [6, 7]. High levels of oxidative stress are mediated by many factors, including abnormal sodium metabolism control in the kidneys, increased activity of angiotensin II, and smoking also occurs in conditions such as hypertension. Reactive oxygen species are produced by multiple enzyme systems found throughout the whole vascular system. Important sources of vascular ROS include the mitochondrial electron transport chain. NADPH oxidases (NOX), xanthine oxidase, and endothelial nitric oxide synthase (eNOS) [8]. All of the above enzymes catalyze the transfer of electrons from their respective substrates to oxygen molecules, reducing oxygen molecules. Basal production of ROS is essential to ensure signaling and cellular homeostasis. The increased concentrations of ROS can be beneficial in some cases, particularly in the oxidative burst of macrophages that is required to eliminate pathogens. Oxidative stress occurs when there is an excess of ROS, whether due to their increased formation and/or a decrease in antioxidant capacity. Increased levels of ROS in the subendothelial space cause oxidation of LDL (low density lipoproteins) particles. This is one of the first steps toward atherosclerosis [9]. Monocytes entering the subendothelial space transform themselves into macrophages, which are then transformed into foam cells under the influence of oxidized cholesterol-rich LDL particles. At the same time, the process of low-grade chronic inflammation is activated. Many cytokines, including interleukin-1 (IL-1), interferon-γ (IFN-γ), tumor necrosis factor-β (TNF-β), and angiotensin II, as well as several chemokines (monocyte chemoattractant protein-1—MCP 1, interleukin-8—IL-8, CXC cytokines and eotaxin) increase the expression of adhesion molecules such as vascular cell adhesion molecule-1 (VCAM-1), P- and E-selectin. Increased activity of adhesion molecules attracts inflammatory and immune cells from the bloodstream (including monocytes, T and B lymphocytes, leukocytes, and mast cells) [10]. Leukocytes bound to adhesion molecules also enter the subendothelial space. At the same time, smooth muscle cells migrate from the intima to the medium and their proliferation and collagen secretion with the formation of an atherosclerotic plaque. These processes are mainly induced by growth factors such as platelet-derived growth factor (PDGF) and transforming growth factor β (TGF-β). These lesions develop through remodeling and neovascularization, which can consequently cause CVD events [11]. The critical component of atherogenesis is dysfunctional endothelium characterized by endothelial leakage, increased production of ROS, secretion of proinflammatory cytokines, increased expression of surface adhesion markers, and decreased production of nitric oxide (NO) [12]. Endothelial dysfunction increases the risk of LDL passing into subendothelial space, where LDL can be oxidated with the participation of ROS. Due to the constitutive expression of eNOS, endothelial cells are the main source of NO. On the other hand, they are also important sources of superoxide and peroxynitrite due to the uncoupling of eNOS in response to BH4 depletion [13]. As showed by Ponnuswamy et al., eNOS deficiency in ApoE−/− mice reduces superoxide production, which indicates that eNOS uncoupling occurs during atherosclerosis [14]. Thus, the endothelium is both a source and a target of ROS. Oxidative stress supports endothelium in a proinflammatory state, which is essential for atherosclerosis. Supplementation with hydrogen peroxide increases the expression of the granule membrane protein, which facilitates the binding of neutrophils to the surface of endothelial cells [15]. Excess ROS and oxidative stress also induce NFκB, which promotes the expression of adhesion molecules such as ICAM-1, VCAM-1, and E-selectin, and cytokines such as TNF-α, as mentioned above [16].

3. The effect of quercetin on atherosclerosis

Quercetin and quercetin glycosides act as strong antioxidants by scavenging free radicals, causing oxidative stress through their phenolic hydroxyls. It seems that the basic mechanism of the antiatherosclerotic effects of quercetin and its glycosides is an indirect antioxidant activity. It is likely that these compounds are involved in modulating the expression and activity of enzymes responsible for the regulation of oxidative stress. They also reduce the inflammatory response to various insults by suppressing the production of proinflammatory substances (e.g., several enzymes and cytokines) [17, 18].

Possible mechanisms of anti-inflammatory action of phenolic compounds include:

- Inhibition of the NFκB signaling pathway. This inhibition leads to downregulation of gene expression of proinflammatory enzymes, e.g., inducible NOS (iNOS), cyclooxygenase-2 (COX-2), and 5-lipoxygenase (5-LOX), and proinflammatory cytokines (IL-1β, IL-6, TNF-α) [19].

- Reduction of monocyte adhesion to endothelial cells by suppressing the expression of E-selectin, MCP-1, ICAM-1, and VCAM-1 [20].
- Reduction of endothelial oxidative stress by inhibition of ROS production dependent on NOX.
- Increasing the levels of the endothelial antiatherosclerotic factor NO by acceleration of eNOS phosphorylation via mechanisms dependent on AMPK (AMP-activated protein kinase).
- Elevation of eNOS expression by activation of Nrf2 E2 (nuclear factor erythroid 2-related factor 2) pathway [21].
- Suppression of vascular smooth muscle cells (VSMC) migration by inhibition of platelet-derived growth factor signaling molecules, such as phosphoinositide-3-kinase (PI3K).
- Suppression of VSMC growth by inducing apoptosis via P38 mitogen-activated protein kinase and p53 signaling pathways activation [22, 23].

A recently published systematic meta-analysis of 16 RCTs (randomized controlled trials) published between 2007 and 2017 focused on the effects of quercetin on the lipid profile in subjects with metabolic syndrome. The authors of the meta-analysis report that quercetin led to significant reductions in total and LDL cholesterol without affecting triglyceride levels. Unfortunately, daily doses and treatment durations used in the trials varied considerably. Treatment duration varied from 3 to 12 weeks, and daily doses from 3.12 to 3000 mg [24]. Another meta-analysis of 9 RCTs conducted in overweight and obese patients confirmed that quercetin supplementation could significantly reduce LDL cholesterol levels at doses ≥250 mg per day and at a total dose ≥14,000 mg (i.e., at least 56 days of quercetin administration) [25]. It is assumed that quercetin also interferes in the activity of matrix metalloproteinases (MMPs). In the studies using molecular modeling techniques, cultured endothelial cells, murine macrophage cells, and in hypertensive rats, quercetin downregulated the expression of MMP-1, MMP-2, and MMP-9. This is the effect that translates into the prevention of plaque instability [26–28]. Quercetin was found to have an antiaggregatory effect on rat platelet-rich plasma depending on the concentration [29]. A synergistic increase in the antiplatelet effect was noted when quercetin was added to aspirin [30]. In addition, isorhamnetin and tamarixetin, two methylated metabolites of quercetin, have been shown to inhibit platelet aggregation and thrombus formation *in vitro* via effects on activation processes such as integrin activation, granule secretion, and intracellular Ca^{2+} mobilization. Their antithrombotic effect was confirmed in mouse cremaster arterioles with laser-induced thrombi [31]. Quercetin significantly increases cyclic adenosine monophosphate (AMP) levels and inhibits arachidonic acid and adenosine diphosphate (ADP)-induced platelet aggregation, in human platelets [32].

4. The effect of quercetin on hypertension

The most common cardiovascular disease present in 1.28 billion adults worldwide is hypertension. Hypertension is a recognized risk factor for the development of atherosclerosis and other cardiovascular diseases. It is estimated that almost 50%

of adults with hypertension are not aware of the elevated blood pressure. There are only 20–30% of treated people for hypertension that are under the control [33]. Bioflavonoids, especially quercetin, one of the most important of these, can find their place in the prevention and treatment of hypertension due to their anti-inflammatory potential, as recent findings suggest [34]. There are many mechanisms of hypertension. All of them result in endothelial dysfunction. In addition, oxidative stress accompanied by ROS contributes to this dysfunction. Consequently, this dysfunction disconnects the endothelial nitric oxide synthase (eNOS) which leads to a reduction of the bioavailability of nitric oxide (NO). NO is an endogenous relaxing factor that regulates vascular tone as well as vascular and cardiac remodelation. Decreased NO levels are associated with hypoxia and the progression of cardiovascular diseases in patients with pre-existing vascular dysfunction [35]. The contribution to vasodilatation of healthy individuals is ambiguous. It is assumed that the protective effect of quercetin on endothelial dysfunction is related to its antihypertensive effect. The scavenging activity of quercetin against ROS helps this effect and reduces endoplasmic stress. *In vitro* study from Lin et al. demonstrated that 20 μmol of quercetin reduced intracellular ROS levels in endothelial cells of mesenteric arteries isolated from hypertensive and normotensive animals [36]. Quercetin can improve vascular function via the activated protein kinase (AMPK) pathway inducing activation of eNOS and, thus, NO production, as ex vivo endothelial function studies have shown [37]. Both *in vitro* studies conducted by Pereira et al. [38] and Lin et al. [36] confirmed the involvement of quercetin-induced autophagy, which leads to improved endothelial cell quality and increased NO production. Long-term administration of quercetin in an animal study induced progressive reduction of SBP (systolic blood pressure) in spontaneously hypertensive rats (SHR). This effect reached statistical significance after the first week of treatment, while in a group of Wistar Kyoto rats, no changes were observed. After 5 weeks of treatment, direct measurements of blood pressure in conscious rats showed that quercetin treatment on SHR induced a significant reduction in systolic (−18%), diastolic (−23%), and mean (−21%) arterial blood pressure. Quercetin also significantly reduced the heart rate (−12%) in these rats [39]. Besides that, administration of quercetin appears to correlate with a reduction in oxidative stress in the aortas of 2K1C rats. However, its effect on systolic blood pressure (SBP) values in 2K1C rats is controversial. A dose-dependent decrease in SHR was observed, with a high dose of quercetin (>7 mg/kg) required to achieve a significant decrease ($p < 0.05$) in both systolic (SBP) and diastolic blood pressure (DBP) [38]. The results of another animal study, where rats were fed a high-salt content diet (0.8% NaCl) during 12 weeks, showed, as expected, an increase in systolic, diastolic, pulse, and mean arterial blood pressure. Chronic salt overload also led to an increase in lipid peroxidation and a decrease in the activity of antioxidant enzymes. Treatment with rutin and quercetin for almost 2 weeks resulted in a remarkable reversal of these indicators when compared with animals that continued only on a high-salt diet (no treatment group). A high-salt content diet also led to a significant increase in concentrations of urea, creatinine, glucose, triacylglycerols, total cholesterol, and low-density lipoproteins. The effect of a high-salt diet on the observed parameters was partially reduced by the administration of rutin or quercetin, while the reference drug nitrendipine showed a smaller effect than these two flavonoids. The results of this study confirm the role of rutin and quercetin as relatively potent antihypertensives and antioxidants [40]. In several *in vitro* and in vivo studies, quercetin has been shown to inhibit angiotensin-converting enzyme (ACE) by binding a zinc molecule to the active site of the ACE. Blocking the active site of the enzyme slows down the conversion of angiotensin I to angiotensin II. This effect is probably related to the chemical structure of flavonoids, mainly the presence of a 30–40 catechol group on the B ring and a double bond and a ketone group on the C ring [41, 42].

On the other hand, the study by Carlstrom et al. did not demonstrate the effect of a diet supplemented with quercetin on the development of arterial hypertension and its complications (e.g., vascular dysfunction and remodeling, hypertrophy of the left ventricular myocardium) in spontaneously hypertensive rats. However, they attributed this discrepancy to the route of administration [43].

Egert et al. found that quercetin in humans led to a significant reduction in systolic blood pressure in all obese and hypertensive participants [44]. On the other hand, Edwards et al. reported that quercetin had no significant effect on blood pressure in prehypertensive patients. Significant reduction in systolic and diastolic blood pressure was shown only in case of patients with stage 1 hypertension [45]. A meta-analysis of data from nine treatment arms of RCTs showed a significant reduction in SBP and DBP after quercetin supplementation. The effect of quercetin probably also depends on the length of its administration. A decrease in both SBD and DBP at the borderline of statistical significance was demonstrated in studies lasting ≥8 weeks, while no significant effect on blood pressure was found in studies in which quercetin was administered for less than 8 weeks. When the studies were categorized by dose of quercetin, there were significant reductions in systolic BP in RCTs with doses ≥500 mg/day and lack of significant effect at doses <500 mg/day. Adjusted indirect comparison did not suggest any significant difference between either of the doses [46].

5. Quercetin and dyslipidemia

Quercetin supplementation significantly reduced levels of plasmatic triglycerides at doses above 50 mg/day, as pointed out in the recent meta-analysis by Sahebkar et al. [47]. However, the available evidence from the 5 RCTs included in this meta-analysis showed a non-significant relationship between quercetin supplementation and other lipid metabolism parameters (total cholesterol, LDL, HDL) [47]. Another systematic review of 9 RCTs described by Guo et al. showed no statistically significant changes in plasma lipids with quercetin supplementation, with the exception of the administration of higher doses of quercetin (more than 250 mg/day) to overweight and obese subjects, in whom it might lower LDL cholesterol [25]. Some other researchers supported the beneficial effects of quercetin in patients with dyslipidemia and metabolic syndrome, when used as add-on therapy. Mazza et al. demonstrated improved plasmatic levels of triglycerides and LDL in hypertensive and dyslipidemic patients with statin intolerance when receiving quercetin in combination with ezetimibe [48]. Quercetin supplementation significantly reduced plasma levels of total cholesterol, LDL, and CRP in patients with metabolic syndrome, as a recent meta-analysis of 16 RCTs has shown [24]. In animal studies, serum triglycerides and total cholesterol levels decreased after quercetin consumption, but this effect was observed only in animals with higher blood lipid concentrations. Quercetin had a beneficial effect on plasma lipid profile in db/db mice [49]. In a study in Western diet-fed mice, Kobori et al. found that quercetin reduced the expression of peroxisome proliferator-activated receptor-α (PPAR-α) and sterol regulatory element-binding protein-1c (SREBP-1c) in the liver, leading to a decrease in triacylglycerol synthesis [50]. Quercetin appears to reduce acetyl-CoA carboxylase (ACC) activity and de novo synthesis of fatty acids and triglycerides in rat hepatocytes. This mechanism could explain the documented effect of quercetin on reducing the concentration of triacylglycerols. Thus, quercetin may improve dyslipidemia through multiple mechanisms, particularly by regulating the expression of PPAR-α, SREBP-1c, and ACC [51].

In addition, quercetin showed a protective effect against high-cholesterol diet-induced cardiac diastolic dysfunction in hyperglycemic rats. This effect is probably caused by the hindrance of cholesterol accumulation and the reduction of ATP, which prevents the change in the expression of PGC-1a, UCP2, and PPARγ receptors [52]. *In vitro* study by Sun et al. in human THP-1-derived macrophage cells showed that quercetin causes cholesterol efflux, reducing foam cell formation. This potentially slows down the development and progression of atherosclerosis. This effect of quercetin appears to be mediated by up-regulation of the cholesterol transporter ABCA1 and the transcription factor PPARγ [53]. Cui et al. reported the same findings after an 8-week-long treatment of apolipoprotein E-deficient atherosclerotic mice with quercetin [54]. Several studies have shown that quercetin can prevent the overexpression of genes for ICAM-1 and MCP-1 and thereby inhibit cell migration in atherosclerotic plaques, reducing the risk of stroke [55]. Liang et al. underlined the importance of developing new delivery systems that are able to increase the solubility of quercetin in their studies in cell lines *in vitro* and in an animal model in vivo [56].

6. Quercetin and dysrhythmias

The energy requirements for sustained electrical and contractile activity of the heart muscle are from more than 90% saturated with cellular adenosine triphosphate (ATP), which is produced by mitochondrial oxidative phosphorylation. Reactive oxygen species are widely present in human cells and, among other things, act as active factors in information transfer and cell apoptosis. As already mentioned, ROS are mainly produced in the mitochondrial electron transport chain (ETC), xanthine oxidase, NADPH oxidase (NOX), and nitric oxide synthase (NOS) [57]. Their excess, which can have a harmful effect, is then eliminated by enzymes such as superoxide dismutase (SOD), catalase, etc. An excess of ROS causes mitochondrial oxidative stress, which negatively affects cardiac excitability, mainly by worsening the function of sodium, potassium, calcium channels, and transporters. This potentially leads to structural and electrical remodeling of the heart with the possible occurrence of various types of arrhythmias [58, 59]. In response to various types of damage, the heart responds adaptively by cardiac remodeling, which is characterized by compensatory maladaptive cardiomyocyte hypertrophy and collagen production with fibrosis formation. Accumulation of the extracellular matrix (EMC) in the myocardium increases the stiffness of the ventricles and impairs relaxation and contraction of the heart. This process, known as cardiac fibrosis, is known to be the underlying pathological basis of various heart diseases. The three main components of arrhythmogenesis and arrhythmia maintenance are:

- substrates,
- triggers,
- and facilitators.

There is ample evidence that fibrosis is one of the main substrates for the development and maintenance of arrhythmias. It is proven that fibrosis can contribute to the development and maintenance of arrhythmias by several mechanisms, primarily through direct electrophysiological mechanisms, as well as indirect cellular mechanisms [60, 61].

Fibrosis can be divided into four different patterns: compact, interstitial, patchy, and diffuse. Of these four types, patchy and interstitial fibrosis in particular can disrupt the electrical connections between cardiomyocytes, causing discontinuous or "zig-zag" conduction that induces arrhythmias. In addition to discontinuous "zig-zag" conduction, cardiac fibrosis may induce other electrophysiological abnormalities through the heterocellular electrical coupling of myofibroblasts and cardiomyocytes, contributing to the development of arrhythmias [62]. Quercetin can positively influence arrhythmias mainly through its effect on cardiac ion channels, improving calcium homeostasis, affecting gap junctions and mitochondrial channels, inhibiting mitochondrial oxidative stress and suppressing cardiac fibrosis, inflammation, modulation of autophagy and apoptosis, as well as improving ischemia/reperfusion injury and gut microbiota [63]. Quercetin also regulates or even directly inhibits critical signaling pathways and key molecules involved in the pathomechanism of arrhythmias, such as TGF-β/Smad, NF-κB and PI3K/AKT and others [64]. In summary, it appears that quercetin can prevent and treat arrhythmias by affecting multiple targets, directions, and pathways in arrhythmogenesis. This may have great potential and clinical application value in the future. Nevertheless, there are some objective limitations of the current research and many questions remain to be resolved.

7. Quercetin in diabetes mellitus

Diabetes mellitus is best defined as a group of endocrine diseases characterized by hyperglycemia as a result of defective insulin secretion or insulin resistance. Insulin signaling has been well defined at the molecular level, but the exact reason for insulin resistance in tissues (especially skeletal muscle), adipose tissue, or liver has not yet been precisely elucidated. Imbalance between ROS and antioxidant defense systems leading to oxidative stress is evident in T2DM pathogenesis [65]. In addition, ROS directly damages mitochondrial DNA, lipids, and proteins and stimulates mitophagy [66]. Mitochondrial dysfunction reduces the catabolism of metabolic substrates, leading to the accumulation of lipid metabolites that contribute to the development of insulin resistance in skeletal muscles, adipocytes, and liver [67]. Hyperglycemia has a harmful effect on the mitochondrial oxygen consumption rate in pancreatic beta cells. That worsens glucose-responsive insulin secretion and stimulates apoptosis. Mitochondrial-targeted ROS-scavenging antioxidants have been shown to reduce lipid peroxidation and improve disease status in animal models of several diseases, including neurodegenerative diseases, diabetic kidney damage, hypertension, Parkinson's disease, and cardiovascular diseases [68]. Excessive accumulation of lipid metabolites inside cells also activates the serine kinase pathway, leading to reduced insulin stimulation and decreased hepatic glucose uptake. Reduced quantity and activity of insulin receptors, serine/threonine hyperphosphorylation of the insulin receptor substrate IRS-1, reduced activity of PI3K, Akt kinase, and PTP1B, and defective expression of GLUT4 have been reported among patients with T2DM [69].

Numerous *in vitro* and prospective animal and clinical studies provide significant evidence that bioflavonoids can be considered as potential agents for the prevention and treatment of diabetes and its complications. Studies conducted on streptozotocin (STZ) induced diabetes rats and type 2 diabetic rats revealed that quercetin can lower blood glucose levels and improve glucose tolerance [70, 71]. Quercetin has also been shown to activate adenosine monophosphate-activated protein kinase (AMPK) in rat liver, which reduces glucose synthesis primarily through the downregulation of glycogen isoenzymes such as phosphoenolpyruvate carboxylase (PEPCK) and glucose-6-phosphatase (G6Pase) [72]. By promoting GLUT4 translocation to cell membranes in mouse skeletal muscle cells, quercetin increases glucose re-uptake. These findings suggest that quercetin may be involved in the regulation of glucose metabolism, increasing glycolysis, and decreasing gluconeogenesis [73].

8. Conclusion

Quercetin is one of the most potent and widespread biologically active flavonoids found in fruits and vegetables. Flavonoids are special chemicals in plants called phytonutrients and have a wide range of health benefits. Diet plays a vital role in reducing the risk of cardiovascular diseases such as heart disease and stroke. Because fruits and vegetables contain flavonoids, eating more of them can reduce the risk of these diseases. As mentioned above, quercetin can improve a wide range of cardiovascular diseases and many risk factors for their development.

In doses commonly found in food, quercetin is unlikely to interact with drugs used for various diseases. The question is whether it affects the effect of other drugs in supranormal doses when supplemented. As mentioned above, quercetin has vasodilating effects. Although this effect may be beneficial in some cases, it may also potentiate the effects of antihypertensive drugs, potentially leading to unwanted hypotension. Quercetin can thus interact with various antihypertensives, such as ACE inhibitors, calcium channel blockers, and beta blockers. Quercetin is found in foods such as onions, apples, and tea and has mild diuretic properties. Thus, when combined with diuretics, the diuretic effects of quercetin may be additive, leading to an increased risk of dehydration or electrolyte imbalance [74, 75]. Therefore, monitoring blood pressure and some biochemical parameters is important when administering quercetin preparations and antihypertensives simultaneously. In addition, some studies suggest that quercetin can inhibit enzymes of the cytochrome P450 system, which, among other things, are involved in drug metabolism. This inhibition could potentially affect the metabolism and clearance of some drugs and thus lead to changes in the levels of these drugs in the body [76, 77].

However, regardless of its mechanism of action, we still need to confirm the cardiovascular risk-reducing effect of quercetin in humans and its potential pharmacological interactions in larger randomized clinical trials.

Author details

Marek Pytliak* and Viliam Vaník
Hospital of the Army Gen. L. Svoboda a.s., Svidnik, Slovakia

*Address all correspondence to: marek.pytliak@svetzdravia.com

References

[1] World Health Organization. Global Action Plan on Physical Activity 2018-2030: More Active People for a Healthier World. World Health Organization; 2019. Available from: https://www.who.int/publications/i/item/9789241514187

[2] WHO CVD Risk Chart Working Group. World Health Organization cardiovascular disease risk charts: Revised models to estimate risk in 21 global regions. The Lancet Global Health. 2019;7:e1332-e1345

[3] Lapuente M, Estruch R, Shahbaz M, Casas R. Relation of fruits and vegetables with major cardiometabolic risk factors, markers of oxidation, and inflammation. Nutrients. 2019;**11**:2381. DOI: 10.3390/nu11102381

[4] Manach C, Morand C, Crespy V, Demigné C, Texier O, Régérat F, et al. Quercetin is recovered in human plasma as conjugated derivatives which retain antioxidant properties. FEBS Letters. 1998;**426**(3):331-336. DOI: 10.1016/s0014-5793(98)00367-6

[5] Kaşıkcı MB, Bağdatlıoğlu N. Bioavailability of quercetin. Current Research in Nutrition and Food Science. 2016;**4**(Special Issue Conference October 2016). Available from: http://www.foodandnutritionjournal.org/?p=2963

[6] Li H, Horke S, Förstermann U. Vascular oxidative stress, nitric oxide and atherosclerosis. Atherosclerosis. 2014;**237**(1):208-219. DOI: 10.1016/j.atherosclerosis.2014.09.001

[7] Xia N, Daiber A, Förstermann U, Li H. Antioxidant effects of resveratrol in the cardiovascular system. British Journal of Pharmacology. 2017;**174**(12):1633-1646. DOI: 10.1111/bph.13492

[8] Snezhkina AV, Kudryavtseva AV, Kardymon OL, Savvateeva MV, Melnikova NV, Krasnov GS, et al. ROS generation and antioxidant defense systems in normal and malignant cells. Oxidative Medicine and Cellular Longevity. 2019;**2019**:6175804. DOI: 10.1155/2019/6175804

[9] Steinberg D, Parthasarathy S, Carew TE, Khoo JC, Witztum JL. Beyond cholesterol. Modifications of low-density lipoprotein that increase its atherogenicity. The New England Journal of Medicine. 1989;**320**(14):915-924. DOI: 10.1056/NEJM198904063201407

[10] Farahi L, Sinha SK, Lusis AJ. Roles of macrophages in atherogenesis. Frontiers in Pharmacology. 2021;**12**:785220. DOI: 10.3389/fphar.2021.785220

[11] Mahdinia E, Shokri N, Taheri AT, et al. Cellular crosstalk in atherosclerotic plaque microenvironment. Cell Communication and Signaling: CCS. 2023;**21**:125. DOI: 10.1186/s12964-023-01153-w

[12] Gimbrone MA Jr, García-Cardeña G. Endothelial cell dysfunction and the pathobiology of atherosclerosis. Circulation Research. 2016;**118**(4):620-636. DOI: 10.1161/CIRCRESAHA.115.306301

[13] Janaszak-Jasiecka A, Płoska A, Wierońska JM, Dobrucki LW, Kalinowski L. Endothelial dysfunction due to eNOS uncoupling: Molecular mechanisms as potential therapeutic targets. Cellular & Molecular Biology Letters. 2023;**28**(1):21. DOI: 10.1186/s11658-023-00423-2

[14] Ponnuswamy P, Schröttle A, Ostermeier E, Grüner S, Huang PL,

Ertl G, et al. eNOS protects from atherosclerosis despite relevant superoxide production by the enzyme in apoE mice. PLoS One. 2012;7(1):e30193. DOI: 10.1371/journal.pone.0030193

[15] Patel KD, Zimmerman GA, Prescott SM, McEver RP, McIntyre TM. Oxygen radicals induce human endothelial cells to express GMP-140 and bind neutrophils. The Journal of Cell Biology. 1991;**112**(4):749-759. DOI: 10.1083/jcb.112.4.749

[16] Zhang H, Park Y, Wu J, Xp C, Lee S, Yang J, et al. Role of TNF-alpha in vascular dysfunction. Clinical Science (London, England). 2009;**116**(3):219-230. DOI: 10.1042/CS20080196

[17] Terao J. Potential role of quercetin glycosides as anti-atherosclerotic food-derived factors for human health. Antioxidants (Basel). 2023;**12**(2):258. DOI: 10.3390/antiox12020258

[18] Qi W, Qi W, Xiong D, Long M. Quercetin: Its antioxidant mechanism, antibacterial properties and potential application in prevention and control of Toxipathy. Molecules. 2022;**27**(19):6545. DOI: 10.3390/molecules27196545

[19] Sun ZJ, Chen G, Hu X, et al. Activation of PI3K/Akt/IKK-α/NF-κB signaling pathway is required for the apoptosis-evasion in human salivary adenoid cystic carcinoma: Its inhibition by quercetin. Apoptosis. 2010;**15**:850-863. DOI: 10.1007/s10495-010-0497-5

[20] Panicker SR, Sreenivas P, Babu MS, Karunagaran D, Kartha CC. Quercetin attenuates monocyte chemoattractant protein-1 gene expression in glucose primed aortic endothelial cells through NF-kappaB and AP-1. Pharmacological Research. 2010;**62**(4):328-336. DOI: 10.1016/j.phrs.2010.06.003

[21] Jia H, Zhang Y, Si X, Jin Y, Jiang D, Dai Z, et al. Quercetin alleviates oxidative damage by activating nuclear factor erythroid 2-related factor 2 signaling in porcine enterocytes. Nutrients. 2021;**13**(2):375. DOI: 10.3390/nu13020375

[22] Endale M, Park SC, Kim S, Kim SH, Yang Y, Cho JY, et al. Quercetin disrupts tyrosine-phosphorylated phosphatidylinositol 3-kinase and myeloid differentiation factor-88 association, and inhibits MAPK/AP-1 and IKK/NF-κB-induced inflammatory mediators production in RAW 264.7 cells. Immunobiology. 2013;**218**(12):1452-1467. DOI: 10.1016/j.imbio.2013.04.019

[23] Kim GT, Lee SH, Kim JI, Kim YM. Quercetin regulates the sestrin 2-AMPK-p38 MAPK signaling pathway and induces apoptosis by increasing the generation of intracellular ROS in a p53-independent manner. International Journal of Molecular Medicine. 2014;**33**(4):863-869. DOI: 10.3892/ijmm.2014.1658

[24] Tabrizi R, Tamtaji OR, Mirhosseini N, Lankarani KB, Akbari M, Heydari ST, et al. The effects of quercetin supplementation on lipid profiles and inflammatory markers among patients with metabolic syndrome and related disorders: A systematic review and meta-analysis of randomized controlled trials. Critical Reviews in Food Science and Nutrition. 2020;**60**(11):1855-1868. DOI: 10.1080/10408398.2019

[25] Guo W, Gong X, Li M. Quercetin actions on lipid profiles in overweight and obese individuals: A systematic review and meta-analysis. Current Pharmaceutical Design. 2019;**25**: 3087-3095. DOI: 10.2174/1381612825666190829153552

[26] Song L, Xu M, Lopes-Virella MF, Huang Y. Quercetin inhibits matrix metalloproteinase-1 expression in human vascular endothelial cells through

extracellular signal-regulated kinase. Archives of Biochemistry and Biophysics. 2001;**391**:72-78. DOI: 10.1006/abbi.2001.2402

[27] Saragusti AC, Ortega MG, Cabrera JL, Estrin DA, Marti MA, Chiabrando GA. Inhibitory effect of quercetin on matrix metalloproteinase 9 activity molecular mechanism and structure-activity relationship of the flavonoid-enzyme interaction. European Journal of Pharmacology. 2010;**644**:138-145. DOI: 10.1016/j.ejphar.2010.07.001

[28] Scoditti E, Calabriso N, Massaro M, Pellegrino M, Storelli C, Martines G, et al. Mediterranean diet polyphenols reduce inflammatory angiogenesis through MMP-9 and COX-2 inhibition in human vascular endothelial cells: A potentially protective mechanism in atherosclerotic vascular disease and cancer. Archives of Biochemistry and Biophysics. 2012;**527**:81-89. DOI: 10.1016/j.abb.2012.05.003

[29] Ko EY, Nile SH, Jung YS, Keum YS. Antioxidant and antiplatelet potential of different methanol fractions and flavonols extracted from onion (*Allium cepa* L.). 3 Biotech. 2018;**8**:155. DOI: 10.1007/s13205-018-1184-4

[30] Stainer AR, Sasikumar P, Bye AP, Unsworth AJ, Holbrook LM, Tindall M, et al. The metabolites of the dietary flavonoid quercetin possess potent antithrombotic activity, and interact with aspirin to enhance antiplatelet effects. TH Open. 2019;**3**:e244-e258. DOI: 10.1055/s-0039-1694028

[31] Bojic M, Debeljak Z, Tomicic M, Medic-Saric M, Tomic S. Evaluation of antiaggregatory activity of flavonoid aglycone series. Nutrition Journal. 2011;**10**:73. DOI: 10.1186/1475-2891-10-73

[32] Boyanov KO, Maneva AI. Influence of platelet aggregation modulators on cyclic amp production in human thrombocytes. Folia Medica. 2018;**60**:241-247. DOI: 10.1515/folmed-2017-0091

[33] NCD Risk Factor Collaboration (NCD-RisC). Worldwide trends in hypertension prevalence and progress in treatment and control from 1990 to 2019: A pooled analysis of 1201 population-representative studies with 104 million participants. Lancet. 2021;**398**(10304):957-980. DOI: 10.1016/S0140-6736(21)01330-1

[34] Xu D, Hu MJ, Wang YQ, Cui YL. Antioxidant activities of quercetin and its complexes for medicinal application. Molecules. 2019;**24**(6):1123. DOI: 10.3390/molecules24061123

[35] Münzel T, Camici GG, Maack C, Bonetti NR, Fuster V, Kovacic JC. Impact of oxidative stress on the heart and vasculature: Part 2 of a 3-part series. Journal of the American College of Cardiology. 2017;**70**(2):212-229. DOI: 10.1016/j.jacc.2017.05.035

[36] Lin X, Han T, Fan Y, Wu S, Wang F, Wang C. Quercetin improves vascular endothelial function through promotion of autophagy in hypertensive rats. Life Sciences. 2020;**258**:118106. DOI: 10.1016/j.lfs.2020.118106

[37] Shen Y, Croft KD, Hodgson JM, Kyle R, Lee IL, Wang Y, et al. Quercetin and its metabolites improve vessel function by inducing eNOS activity via phosphorylation of AMPK. Biochemical Pharmacology. 2012;**84**(8):1036-1044. DOI: 10.1016/j.bcp.2012.07.016

[38] Pereira SC, Parente JM, Belo VA, Mendes AS, Gonzaga NA, do Vale GT, et al. Quercetin decreases the activity of matrix metalloproteinase-2 and ameliorates vascular remodeling in renovascular hypertension.

Atherosclerosis. 2018;**270**: 146-153. DOI: 10.1016/j. atherosclerosis.2018.01.031

[39] Duarte J, Pérez-Palencia R, Vargas F, Ocete MA, Pérez-Vizcaino F, Zarzuelo A, et al. Antihypertensive effects of the flavonoid quercetin in spontaneously hypertensive rats. British Journal of Pharmacology. 2001;**133**(1):117-124. DOI: 10.1038/sj.bjp.0704064

[40] Olaleye MT, Crown OO, Akinmoladun AC, Akindahunsi AA. Rutin and quercetin show greater efficacy than nifedipin in ameliorating hemodynamic, redox, and metabolite imbalances in sodium chloride-induced hypertensive rats. Human & Experimental Toxicology. 2014;**33**(6):602-608. DOI: 10.1177/0960327113504790

[41] Häckl LP, Cuttle G, Dovichi SS, Lima-Landman MT, Nicolau M. Inhibition of angiotesin-converting enzyme by quercetin alters the vascular response to brandykinin and angiotensin I. Pharmacology. 2002;**65**(4):182-186. DOI: 10.1159/000064341

[42] Loizzo MR, Said A, Tundis R, Rashed K, Statti GA, Hufner A, et al. Inhibition of angiotensin converting enzyme (ACE) by flavonoids isolated from Ailanthus excelsa (Roxb) (Simaroubaceae). Phytotherapy Research. 2007;**21**(1):32-36. DOI: 10.1002/ptr.2008

[43] Carlstrom J, Symons JD, Wu TC, Bruno RS, Litwin SE, Jalili T. A quercetin supplemented diet does not prevent cardiovascular complications in spontaneously hypertensive rats. The Journal of Nutrition. 2007;**137**(3):628-633. DOI: 10.1093/jn/137.3.628

[44] Egert S, Bosy-Westphal A, Seiberl J, Kürbitz C, Settler U, Plachta-Danielzik S, et al. Quercetin reduces systolic blood pressure and plasma oxidised low-density lipoprotein concentrations in overweight subjects with a high-cardiovascular disease risk phenotype: A double-blinded, placebo-controlled cross-over study. The British Journal of Nutrition. 2009;**102**(7):1065-1074. DOI: 10.1017/S0007114509359127

[45] Edwards RL, Lyon T, Litwin SE, Rabovsky A, Symons JD, Jalili T. Quercetin reduces blood pressure in hypertensive subjects. The Journal of Nutrition. 2007;**137**(11):2405-2411. DOI: 10.1093/jn/137.11.2405

[46] Serban MC, Sahebkar A, Zanchetti A, Mikhailidis DP, Howard G, Antal D, et al. Effects of quercetin on blood pressure: A systematic review and meta-analysis of randomized controlled trials. Journal of the American Heart Association. 2016;**5**:e002713. DOI: 10.1161/JAHA.115.002713

[47] Sahebkar A. Effects of quercetin supplementation on lipid profile: A systematic review and meta-analysis of randomized controlled trials. Critical Reviews in Food Science and Nutrition. 2017;**57**(4):666-676. DOI: 10.1080/10408398.2014.948609

[48] Mazza A, Nicoletti M, Lenti S, Torin G, Rigatelli G, Pellizzato M, et al. Effectiveness and safety of novel nutraceutical formulation added to ezetimibe in statin-intolerant hypercholesterolemic subjects with moderate-to-high cardiovascular risk. Journal of Medicinal Food. 2021;**24**(1):59-66. DOI: 10.1089/jmf.2020.0019

[49] Jiang X, Yu J, Wang X, Ge J, Li N. Quercetin improves lipid metabolism via SCAP-SREBP2-LDLr signaling pathway in early stage diabetic nephropathy. Diabetes, Metabolic Syndrome and Obesity. 2019;**12**:827-839. DOI: 10.2147/DMSO.S195456

[50] Kobori M, Masumoto S, Akimoto Y, Oike H. Chronic dietary intake of quercetin alleviates hepatic fat accumulation associated with consumption of a Western-style diet in C57/BL6J mice. Molecular Nutrition & Food Research. 2011;**55**(4):530-540. DOI: 10.1002/mnfr.201000392

[51] Gnoni GV, Paglialonga G, Siculella L. Quercetin inhibits fatty acid and triacylglycerol synthesis in rat-liver cells. European Journal of Clinical Investigation. 2009;**39**(9):761-768. DOI: 10.1111/j.1365-2362.2009.02167.x

[52] Castillo RL, Herrera EA, Gonzalez-Candia A, Reyes-Farias M, de la Jara N, Peña JP, et al. Quercetin prevents diastolic dysfunction induced by a high-cholesterol diet: Role of oxidative stress and bioenergetics in hyperglycemic rats. Oxidative Medicine and Cellular Longevity. 2018;**2018**:7239123. DOI: 10.1155/2018/7239123

[53] Sun L, Li E, Wang F, Wang T, Qin Z, Niu S, et al. Quercetin increases macrophage cholesterol efflux to inhibit foam cell formation through activating PPARγ-ABCA1 pathway. International Journal of Clinical and Experimental Pathology. 2015;**8**(9):10854-10860

[54] Cui Y, Hou P, Li F, Liu Q, Qin S, Zhou G, et al. Quercetin improves macrophage reverse cholesterol transport in apolipoprotein E-deficient mice fed a high-fat diet. Lipids in Health and Disease. 2017;**16**(1):9. DOI: 10.1186/s12944-016-0393-2

[55] Huwait EA, Saddeek SY, Al-MassabiRF, AlmowalladSJ, PushparajPN, Kalamegam G. Antiatherogenic effects of quercetin in the THP-1 macrophage model in vitro, with insights into its signaling mechanisms using in silico analysis. Frontiers in Pharmacology. 2021;**12**:698138. DOI: 10.3389/fphar.2021.698138

[56] Liang N, Li YM, He Z, Hao W, Zhao Y, Liu J, et al. Rutin and quercetin decrease cholesterol in HepG2 cells but not plasma cholesterol in hamsters by oral administration. Molecules. 2021;**26**(12):3766. DOI: 10.3390/molecules26123766

[57] Zhao RZ, Jiang S, Zhang L, Yu ZB. Mitochondrial electron transport chain, ROS generation and uncoupling (review). International Journal of Molecular Medicine. 2019;**44**(1):3-15. DOI: 10.3892/ijmm.2019.4188

[58] Phull AR, Nasir B, Haq IU, Kim SJ. Oxidative stress, consequences and ROS mediated cellular signaling in rheumatoid arthritis. Chemico-Biological Interactions. 2018;**281**:121-136. DOI: 10.1016/j.cbi.2017.12.024

[59] Deng J, Jiang Y, Chen ZB, Rhee JW, Deng Y, Wang ZV. Mitochondrial dysfunction in cardiac arrhythmias. Cells. 2023;**12**(5):679. DOI: 10.3390/cells12050679

[60] Purnomo Y, Piccart Y, Coenen T, Prihadi JS, Lijnen PJ. Oxidative stress and transforming growth factor-β1-induced cardiac fibrosis. Cardiovascular & Hematological Disorders Drug Targets. 2013;**13**(2):165-172. DOI: 10.2174/1871529x11313020010

[61] He X, Gao X, Peng L, Wang S, Zhu Y, Ma H, et al. Atrial fibrillation induces myocardial fibrosis through angiotensin II type 1 receptor-specific Arkadia-mediated downregulation of Smad7. Circulation Research. 2011;**108**(2):164-175. DOI: 10.1161/CIRCRESAHA.110.234369

[62] Nguyen MN, Kiriazis H, Gao XM, Du XJ. Cardiac fibrosis and arrhythmogenesis. Comprehensive Physiology. 2017;**7**(3):1009-1049. DOI: 10.1002/cphy.c160046

[63] Zhou Y, Suo W, Zhang X, Lv J, Liu Z, Liu R. Roles and mechanisms of quercetin on cardiac arrhythmia: A review. Biomedicine & Pharmacotherapy. 2022;**153**:113447. DOI: 10.1016/j.biopha.2022.113447

[64] Bertaud A, Joshkon A, Heim X, Bachelier R, Bardin N, Leroyer AS, et al. Signaling pathways and potential therapeutic strategies in cardiac fibrosis. International Journal of Molecular Sciences. 2023;**24**(2):1756. DOI: 10.3390/ijms24021756

[65] Evans JL, Goldfine ID, Maddux BA, Grodsky GM. Are oxidative stress-activated signaling pathways mediators of insulin resistance and beta-cell dysfunction? Diabetes. 2003;**52**(1):1-8. DOI: 10.2337/diabetes.52.1.1

[66] Tangvarasittichai S. Oxidative stress, insulin resistance, dyslipidemia and type 2 diabetes mellitus. World Journal of Diabetes. 2015;**6**(3):456-480. DOI: 10.4239/wjd.v6.i3.456

[67] Kelley DE, He J, Menshikova EV, Ritov VB. Dysfunction of mitochondria in human skeletal muscle in type 2 diabetes. Diabetes. 2002;**51**(10):2944-2950. DOI: 10.2337/diabetes.51.10.2944

[68] Samuel VT, Liu ZX, Qu X, Elder BD, Bilz S, Befroy D, et al. Mechanism of hepatic insulin resistance in non-alcoholic fatty liver disease. The Journal of Biological Chemistry. 2004;**279**(31):32345-32353. DOI: 10.1074/jbc.M313478200

[69] Aguirre V, Uchida T, Yenush L, Davis R, White MF. The c-Jun NH(2)-terminal kinase promotes insulin resistance during association with insulin receptor substrate-1 and phosphorylation of Ser(307). The Journal of Biological Chemistry. 2000;**275**(12):9047-9054. DOI: 10.1074/jbc.275.12.9047

[70] Yang DK, Kang HS. Anti-diabetic effect of cotreatment with quercetin and resveratrol in streptozotocin-induced diabetic rats. Biomolecules and Therapeutics. 2018;**26**(2):130-138. DOI: 10.4062/biomolther.2017.254

[71] Chen S, Jiang H, Wu X, Fang J. Therapeutic effects of quercetin on inflammation, obesity, and type 2 diabetes. Mediators of Inflammation. 2016;**2016**:9340637. DOI: 10.1155/2016/9340637

[72] Zhang F, Feng J, Zhang J, Kang X, Qian D. Quercetin modulates AMPK/SIRT1/NF-κB signaling to inhibit inflammatory/oxidative stress responses in diabetic high fat diet-induced atherosclerosis in the rat carotid artery. Experimental and Therapeutic Medicine. 2020;**20**(6):280. DOI: 10.3892/etm.2020.9410

[73] Jiang H, Yamashita Y, Nakamura A, Croft K, Ashida H. Quercetin and its metabolite isorhamnetin promote glucose uptake through different signalling pathways in myotubes. Scientific Reports. 2019;**9**(1):2690. DOI: 10.1038/s41598-019-38711-7

[74] Larson A, Witman MA, Guo Y, Ives S, Richardson RS, Bruno RS, et al. Acute, quercetin-induced reductions in blood pressure in hypertensive individuals are not secondary to lower plasma angiotensin-converting enzyme activity or endothelin-1: Nitric oxide. Nutrition Research. 2012;**32**(8):557-564. DOI: 10.1016/j.nutres.2012.06.018

[75] Andres S, Pevny S, Ziegenhagen R, Bakhiya N, Schäfer B, Hirsch-Ernst KI, et al. Safety aspects of the use of quercetin as a dietary supplement. Molecular Nutrition & Food Research. 2018;**62**(1). DOI: 10.1002/mnfr.201700447. Epub 2017 Dec 19

[76] Chen Y, Xiao P, Ou-Yang D-S, Fan L, Guo D, Wang Y-N, et al. Simultaneous

action of the flavonoid quercetin on cytochrome P450 (CYP) 1A2, CYP2A6, N-acetyltransferase and xanthine oxidase activity in healthy volunteers. Clinical and Experimental Pharmacology & Physiology. 2009;**36**:828-833. DOI: 10.1111/j.1440-1681.2009.05158.x

[77] Elbarbry F, Ung A, Abdelkawy K. Studying the inhibitory effect of quercetin and thymoquinone on human cytochrome P450 enzyme activities. Pharmacognosy Magazine. 2018;**13**:S895-S899. DOI: 10.4103/0973-1296.224342